P9-AEZ-618

The Chemistry of
Low-Rank Coals

ACS SYMPOSIUM SERIES 264

The Chemistry of Low-Rank Coals

Harold H. Schobert, EDITOR
University of North Dakota

Based on a symposium sponsored by
the Division of Fuel Chemistry
and the Division of Geochemistry
at the 186th Meeting
of the American Chemical Society,
Washington, D.C.,
August 28–September 2, 1983

American Chemical Society, Washington, D.C. 1984

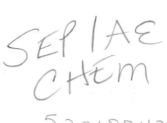

SEP |AE
CHEM

5 2 0 I 8 8 4 2

Library of Congress Cataloging in Publication Data

The chemistry of low-rank coals.
(ACS symposium series, ISSN 0097–6156; 264)

"Based on a symposium jointly sponsored by the
Division of Fuel Chemistry and the Division of
Geochemistry at the 186th Meeting of the American
Chemical Society, Washington, D.C., August
28–September 2, 1983."

Bibliography: p.
Includes indexes.

1. Lignite—Analysis—Congresses.

I. Schobert, Harold H., 1943–   . II. American
Chemical Society. Division of Fuel Chemistry.
III. American Chemical Society. Division of
Geochemistry. IV. Series.

TP329.C53   1984   662.6′222   84–14519
ISBN 0–8412–0866–2

Copyright © 1984

American Chemical Society

All Rights Reserved. The appearance of the code at the bottom of the first page of each
chapter in this volume indicates the copyright owner's consent that reprographic copies of the
chapter may be made for personal or internal use or for the personal or internal use of specific
clients. This consent is given on the condition, however, that the copier pay the stated per
copy fee through the Copyright Clearance Center, Inc., 21 Congress Street, Salem, MA 01970,
for copying beyond that permitted by Sections 107 or 108 of the U.S. Copyright Law. This
consent does not extend to copying or transmission by any means—graphic or electronic—for
any other purpose, such as for general distribution, for advertising or promotional purposes,
for creating a new collective work, for resale, or for information storage and retrieval systems.
The copying fee for each chapter is indicated in the code at the bottom of the first page of the
chapter.

The citation of trade names and/or names of manufacturers in this publication is not to be
construed as an endorsement or as approval by ACS of the commercial products or services
referenced herein; nor should the mere reference herein to any drawing, specification, chemical
process, or other data be regarded as a license or as a conveyance of any right or permission,
to the holder, reader, or any other person or corporation, to manufacture, reproduce, use, or
sell any patented invention or copyrighted work that may in any way be related thereto.
Registered names, trademarks, etc., used in this publication, even without specific indication
thereof, are not to be considered unprotected by law.

PRINTED IN THE UNITED STATES OF AMERICA

UNIV. OF CALIFORNIA
WITHDRAWN

TP329
C53
1984
CHEM

ACS Symposium Series

M. Joan Comstock, *Series Editor*

Advisory Board

Robert Baker
U.S. Geological Survey

Martin L. Gorbaty
Exxon Research and Engineering Co.

Herbert D. Kaesz
University of California— Los Angeles

Rudolph J. Marcus
Office of Naval Research

Marvin Margoshes
Technicon Instruments Corporation

Donald E. Moreland
USDA, Agricultural Research Service

W. H. Norton
J. T. Baker Chemical Company

Robert Ory
USDA, Southern Regional
    Research Center

Geoffrey D. Parfitt
Carnegie–Mellon University

Theodore Provder
Glidden Coatings and Resins

James C. Randall
Phillips Petroleum Company

Charles N. Satterfield
Massachusetts Institute of Technology

Dennis Schuetzle
Ford Motor Company
    Research Laboratory

Davis L. Temple, Jr.
Mead Johnson

Charles S. Tuesday
General Motors Research Laboratory

C. Grant Willson
IBM Research Department

UNIV OF CALIFORNIA
WITHDRAWN

FOREWORD

The ACS SYMPOSIUM SERIES was founded in 1974 to provide a medium for publishing symposia quickly in book form. The format of the Series parallels that of the continuing ADVANCES IN CHEMISTRY SERIES except that in order to save time the papers are not typeset but are reproduced as they are submitted by the authors in camera-ready form. Papers are reviewed under the supervision of the Editors with the assistance of the Series Advisory Board and are selected to maintain the integrity of the symposia; however, verbatim reproductions of previously published papers are not accepted. Both reviews and reports of research are acceptable since symposia may embrace both types of presentation.

CONTENTS

Preface..ix

OVERVIEW

1. **The Chemical Characteristics of Victorian Brown Coal** 3
   G. J. Perry, D. J. Allardice, and L. T. Kiss

2. **Correlations Between Petrographical Properties, Chemical Structure, and Technological Behavior of Rhenish Brown Coal** 15
   E. A. Wolfrum

3. **The Structure and Reactions of Northern Great Plains Lignites** 39
   W. R. Kube, Harold H. Schobert, S. A. Benson, and F. R. Karner

4. **Resources, Properties, and Utilization of Texas Lignite: A Review** 53
   T. F. Edgar and W. R. Kaiser

PHYSICAL AND CHEMICAL CHARACTERIZATION

5. **Small-Angle X-Ray Scattering of the Submicroscopic Porosity of Some Low-Rank Coals** 79
   Harold D. Bale, Marvin L. Carlson, Mohanan Kalliat, Chul Y. Kwak, and Paul W. Schmidt

6. **Determination of the Microstructure of Wet and Dry Brown Coal by Small-Angle X-Ray Scattering** 95
   M. Setek, I. K. Snook, and H. K. Wagenfeld

7. **Chemical Variation as a Function of Lithotype and Depth in Victorian Brown Coal**...109
   R. B. Johns, A. L. Chaffee, and T. V. Verheyen

8. **Some Structural Features of a Wilcox Lignite** 133
   Narayani Mallya and Ralph A. Zingaro

9. **Comparison of Hydrocarbons Extracted from Seven Coals by Capillary Gas Chromatography and Gas Chromatography–Mass Spectrometry** 145
   Sylvia A. Farnum, Ronald C. Timpe, David J. Miller, and Bruce W. Farnum

10. **Analysis of the Inorganic Constituents in Low-Rank Coals** 159
    G. P. Huffman and F. E. Huggins

11. **Geochemical Variation of Inorganic Constituents in a North Dakota Lignite** 175
    F. R. Karner, S. A. Benson, Harold H. Schobert, and R. G. Roaldson

12. **Measurement and Prediction of Low-Rank Coal Slag Viscosity** 195
    Robert C. Streeter, Erle K. Diehl, and Harold H. Schobert

ASPECTS OF REACTIVITY

13. **Role of Exchangeable Cations in the Rapid Pyrolysis of Lignites** 213
    Mark E. Morgan and Robert G. Jenkins

14. **Low-Rank Coal Hydropyrolysis**.....................................227
    C. S. Wen and T. P. Kobylinski

15. **Combustion Reactivity of Chars from Australian Subbituminous Coals** ...243
    B. C. Young

16. **Cationic Effects During Lignite Pyrolysis and Combustion** 255
    Bruce A. Morgan and Alan W. Scaroni

17. **Catalysis of Lignite Char Gasification by Exchangeable Calcium
    and Magnesium** 267
    T. D. Hengel and P. L. Walker, Jr.

18. **Mechanistic Studies on the Hydroliquefaction of Victorian Brown Coal** ...275
    F. P. Larkins, W. R. Jackson, D. Rash, P. A. Hertan, P. J. Cassidy,
    M. Marshall, and I. D. Watkins

19. **Structure and Liquefaction Reactions of Texas Lignite**..................287
    C. V. Philip, R. G. Anthony, and Zhi-Dong Cui

Author Index 305

Subject Index 305

PREFACE

Enormous quantities of low-rank coals—lignite and subbituminous—exist in the United States. The estimated reserve base economically recoverable by underground and strip mining is approximately 230 billion metric tons, an amount nearly equal to the reserve base of bituminous coal. When expressed on an energy basis, to account for the lower average heating value of low-rank coals, the low-rank reserves represent about 45% of the total economically recoverable energy from coal. Low-rank coals are, therefore, a resource of significant potential impact on our national energy economy. They also occur in many other nations, particularly Canada, Australia, several of the central and eastern European countries, and the Soviet Union.

Several characteristics of low-rank coals are sufficiently distinctive to serve as rule-of-thumb discriminators between low-rank coals and bituminous coals. Low-rank coals have much higher as-mined moisture content, which can in some cases exceed 40%. Low-rank coals have a greater content of organic oxygen, particularly in the form of carboxylic acid groups. The inorganic species in low-rank coals contain a higher proportion of alkali and alkaline earth elements and a lower proportion of silicon and aluminum than the mineral matter in bituminous coals. Furthermore, much of the alkali and alkaline earth content is present as mobile cations associated with the organic functional groups rather than as discrete mineral phases. These characteristics affect both laboratory studies and utilization of low-rank coals in unique ways and make it very risky to extrapolate the knowledge derived from bituminous coals to low-rank coals.

Despite the large reserve base and the distinctive characteristics of low-rank coals, until recently—about the mid-1970s—very little research in the United States was specifically devoted to understanding the properties and behavior of low-rank coals. In that respect, we have probably lagged behind the brown-coal research communities in Australia and Europe. The inattention to low-rank coals reflects the historical pattern of coal consumption in the United States. Prior to the energy crisis and the emerging environmental concerns of the 1970s, the use of low-rank coals was confined to local or regional markets of low demand. These markets amounted to about 1% of the total U.S. coal consumption. However, since the mid-1970s a steady increase in low-rank coal utilization has occurred, and with it has come an increasingly vigorous research program dedicated to low-rank coals.

The selection of papers for this book was guided by two considerations. First, when the symposium from which this book originated was presented, a collection of papers dealing with current fundamental research on low-rank coals could not be found. On the other hand, the technology and utilization of low-rank coals on pilot-plant or commercial scale had been well served by the biennial Lignite Symposia, the most recent of which was also held in 1983. Second, the American low-rank coal research community could learn much from colleagues in Australia and Europe. Consequently, in recruiting and selecting papers for the symposium, particular emphasis was given to basic research rather than large-scale work, and papers from overseas were especially encouraged.

The chapters are arranged into three groups. The first group provides an overview of four major low-rank coal deposits—Australian and German brown coals, and Gulf Coast and northern Great Plains lignites in the United States. The second treats current research on the physical and chemical characterization of low-rank coals. The third group is a collection of chapters dealing with fundamental studies of low-rank coal reactivity. Where appropriate papers were available, chapters from the United States and from overseas on similar topics are arranged back-to-back.

I thank all of the authors for their efforts in preparing a preprint for the symposium and in subsequently revising the preprint into chapter form while accommodating the recommendations of the peer reviewers and the editor. Particular thanks are due to Tom Edgar and Bill Kaiser for preparing the overview chapter on Texas lignite (which was not a paper in the symposium) under forced draft with a very tight schedule.

Finally, it is a pleasure to acknowledge, on behalf of the Division of Fuel Chemistry, the generous assistance of the Petroleum Research Fund in providing a travel grant that made it possible for several of our overseas colleagues to attend the symposium.

HAROLD H. SCHOBERT
University of North Dakota
Energy Research Center
Grand Forks, ND 58202

OVERVIEW

The Chemical Characteristics of Victorian Brown Coal

G. J. PERRY[1], D. J. ALLARDICE[1], and L. T. KISS[2]

[1]Victorian Brown Coal Council, 136 Exhibition Street, Melbourne 3000, Australia
[2]State Electricity Commission of Victoria, Herman Research Laboratory, Howard Street, Richmond 3121, Australia

Extensive deposits of soft brown coal exist in Tertiary age sediments in a number of areas in Victoria and the largest single deposit occurs in the Latrobe Valley, about 150 kilometers east of Melbourne. In this region the coal seams often exceed 150 metres in thickness, with an overburden to coal ratio usually better than 1:2 making the coal ideally suited for large-scale open-cut mining.

A recent study (1) has estimated the State's brown coal resources to be almost 200,000 million metric tons with approximately 52,000 million metric tons defined as usable reserves. About 85% of this coal is located in the Latrobe Valley.

Since 1920 Latrobe Valley brown coal has been developed for power generation. The State Electricity Commission of Victoria (SECV) wins coal from two major open cuts at Yallourn and Morwell and operates coal fired power stations which presently consume approximately 35 million metric tons per annum. In addition to power generation, small quantities of brown coal are used for briquette manufacture and char production.

Brown coal accounts for about 95% of Victoria's non-renewable energy reserves and it is now recognized that with suitable up-grading, primarily drying, it has the potential to become the basis of the supply of energy in a variety of forms. Currently various studies for major conversion projects proposed by Australian, Japanese and German interests are being undertaken with the co-operation of the Victorian Brown Coal Council, the most advanced project being a 50 metric tons per day hydrogenation pilot plant currently under construction at Morwell, funded by New Energy Development Organization (NEDO) of Japan.

The chemical characteristics of Latrobe Valley brown coals have been extensively studied over the last twenty-five years, primarily in relation to the effect of coal quality on combustion for power generation. More recently a research project was initiated with the objective of determining the characteristics and suitability of the State's brown coal resources for uses other than power generation, primarily conversion to liquid fuels.

0097-6156/84/0264-0003$06.00/0
© 1984 American Chemical Society

This paper outlines the chemical characteristics of Victorian brown coal and discusses the variability of the coal, both between fields and within a seam. The importance of chemical properties in relation to coal quality and the implications for utilization are also briefly addressed.

Properties of Victorian Brown Coal

The development and adaptation of modern analytical techniques for analysis of Victorian brown coal was pioneered jointly in the 1960's by the Commonwealth Scientific and Industrial Research Organization and the State Electricity Commission of Victoria. As a result, the total coal analysis time was halved and the determination of the ash forming constituents directly on the coal took one sixth of the time of conventional ash analysis. More importantly brown coal analysis was put onto a rational basis taking its unique properties into account, thereby providing more pertinent information concerning the genesis, occurrence and use of Victorian brown coal.

Moisture. One of the most important chemical measurements made on brown coal is the bed moisture content which is also a good measure of physical rank; the greater the degree of compaction of the coal and its degree of coalification, the lower is the moisture content. To obtain meaningful results, the sampling and sample preparation have to be carried out quickly to avoid moisture loss. The choice of the method of determination is important as thermal decomposition of functional groups can result in loss of CO_2 as well as H_2O. The preferred methods therefore involve direct measurement of the water released either by azeotropic distillation or adsorption from an intert carrier gas rather than by weight loss of the coal.
    It is important to realize that the bed moisture content of soft brown coals is significantly higher than the equilibrium moisture holding capacity, a parameter which is used to characterize higher rank coals. This is illustrated in Table 1 for a range of Victorian brown coals.
    In terms of moisture the economic value of high-rank coals is best indicated by the moisture holding capacity because it reflects the condition of the coal for utilization. In the case of Victorian brown coal, the bed moisture content is the critical value since the coal is used directly from the open cut.
    Mineral and Inorganic Content. Ash content has been traditionally used to assess the magnitude of combustion residue and to derive the so-called "coal substance" by difference which allows meaningful comparisons of different coals. In this context the ash is used as an approximation of the mineral matter content. The tacit assumption made, of course, is that the ash is derived solely from coal minerals, and this is certainly not the case for Victorian brown coal where the bulk of the ash forming material occurs as inherent inorganic matter in the form of exchangeable cations, associated with oxygen containing functional groups. This is also the case with many other low-rank coals.
    Using a combination of X-ray fluorescence (XRF) on coal pellets and atomic absorption (AA) techniques on acid extracts,

direct chemical analyses of the ash forming elements in Victorian brown coal has been performed. Arising from this, a method of expression of results has been developed (2) which is based on classifying the mineral matter in brown coal into mineral and inorganic matter fractions and expressing each in a way which reflects their occurrence in the coal.

The Inorganics are a group of exchangeable cations and water soluble salts, analyzed by AA on dilute acid extracts from the coal and expressed in terms of chemical analysis on a coal basis as -

$$Inorganics = Na + Ca + Mg + Fe + (Al) + (Si) + NaCl$$

where Fe refers to the non-pyritic iron and (Al) and (Si) to the acid soluble aluminium and silicon respectively. This expression is complicated by the fact that some iron and aluminium can be present as acid soluble hydroxides, but these are not usually significant.

The group named Minerals, which occurs as discrete particles principally of quartz, kaolinite and pyrite/marcasite is expressed in terms of chemical analysis as -

$$Minerals = SiO_2 + Al_2O_3 + TiO_2 + K_2O + FeS_2$$

This expression ignores the water of constitution of clays which is usually of negligible magnitude for Victorian brown coals.

The total weight of Minerals and Inorganics expressed on a dry coal basis gives the best estimate corresponding to "mineral matter" in high-rank coal technology. In the case of Latrobe Valley coals the Inorganics are far more important than Minerals both in quantity and from a utilization point of view.

From a knowledge of the chemical constitution of the mineral matter, it is possible to calculate and predict the composition of the ash or inorganic residue remaining after most technological processes. Table 2 illustrates the comparison between mineral matter content and ash content and the successful calculation of ash content from mineral matter data for a number of typical Victorian brown coals. Table 3 illustrates the futility of predicting the quantities of ash produced in modern, pulverized fuel fired power stations from the empirical ash test results. It illustrates for Morwell and Yallourn coals the quantities of ash produced in a boiler as opposed to the ash test; and it compares the laboratory ash composition with the precipitator ash actually produced. The mineral matter composition is also given as a guide. Both the laboratory and precipitator ash compositions can be calculated from the composition of the mineral matter by allowing for the difference in the degree of sulphation of the Inorganics. By extending this approach it should also be possible to calculate the composition of the inorganic residue obtained in hydrogenation processes.

It should also be noted that oxygen can only be estimated by difference if the mineral matter is known; using the ash value will yield misleading results.

TABLE 1.   BED MOISTURE AND MOISTURE HOLDING CAPACITY FOR SELECTED
           VICTORIAN COALS

Coal Field (Selected Values)	Bed Moisture (As Received)		Moisture Holding Capacity % (Equilibrium Moisture)*
	Kg/Kg Dry Coal	Weight Percent	
Yallourn-Maryvale	1.82	64.2	38.4
Morwell-Narracan	1.54	60.5	38.2
Loy Yang	1.63	61.9	42.1
Flynn	1.76	63.5	41.2
Yallourn North Ext	0.98	49.5	41.2
Coolungoolun	1.19	54.4	35.0
Gormandale	1.27	56.0	39.0
Gelliondale	1.83	64.6	35.3
Stradbroke	1.41	58.4	37.1
Anglesea	0.87	46.6	32.9
Bacchus Marsh	1.53	60.4	31.2

* Determined at 97% humidity and 30°C

TABLE 2.   COMPARISON OF ASH CONTENT AND MINERAL MATTER FOR
           SELECTED VICTORIAN BROWN COALS

Coal Field (Selected Values)	Ash*		Mineral Matter*
	Determined	Calculated	
Yallourn-Maryvale	2.3	2.4	1.7
Morwell-Narracan	2.4	2.6	1.5
Loy Yang	1.0	1.1	0.8
Flynn	1.4	1.5	0.1
Yallourn North Ext	4.3	4.5	3.0
Coolungoolun	2.3	2.4	2.2
Gormandale	1.6	1.6	1.3
Gelliondale	5.8	5.8	3.4
Stradbroke	3.9	3.6	2.0
Anglesea	3.9	3.4	2.2
Bacchus Marsh	7.4	7.4	3.5

* Dry Coal Basis

TABLE 3. COAL-ASH CHEMISTRY : MORWELL OPEN CUT

Coal % Dry Basis			Ash%	
			Laboratory	Boiler Precipitator
Minerals				
FeS_2	0.13	–	–	–
SiO_2	0.19	SiO_2	6.2	8.0
Al_2O_3	0.04	Al_2O_3	1.3	1.7
Inorganics				
Fe	0.32	Fe_2O_3	17.6	22.7
Ca	0.70	CaO	32.0	41.2
Mg	0.24	MgO	13.1	16.8
Na	0.08	Na_2O	3.6	4.6
Cl	0.04	–	–	–
Organic S	0.25	SO_3	26.1	5.0
Total Ash			3.1	2.6

Oxygen Functional Groups. Oxygen is one of the major elements present in the organic substance of Victorian brown coal. For Latrobe Valley brown coals oxygen generally comprises over 25% on a dry mineral and inorganic free (dmif) basis and about half of this oxygen can be accounted for in the acidic functional groups - phenolic hydroxyl, free carboxylic acid and carboxylate. The 40-50% of the oxygen not accounted for as acidic oxygen is primarily contained in carbonyl groups, ether linkages and heterocyclic ring structures.

Variation of Chemical Properties Within a Seam

The variation of chemical properties in a brown coal seam is attributable to the influence of two independent variables, namely coal rank and coal type. Rank variations are due to the burial history of the coal, that is, the time, temperature and pressure it has undergone since its deposition. However, type (or lithotype) variations also significantly influence brown coal properties. Lithotypes arise from variations in the prevailing botanical communities, the depth and nature of the swamp water and in the conditions of decay and decomposition of plant material. In Victorian brown coal these lithotypes are macroscopically recognizable bands or layers within a coal seam which become readily apparent on partially dried and weathered faces of open cuts. The basic factors on which lithotypes are classified are color and texture in air-dried coal, with degree of gelification, weathering pattern and physical properties used as supplementary characteristics which varies from pale to dark rown. In the case of Latrobe Valley coals, the color of the air dried coal as measured by its diffuse reflectance and expressed as a Color Index gives a numerical value related to lithotype.

The variation in rank with depth in the Morwell Open Cut is illustrated by the yearly weighted averages for each operating level. The gradation in rank is clearly illustrated in Figure 1 by the increase in carbon content, and the associated increase in gross dry specific energy; the volatile matter also decreases slightly with depth. These samples were not selected on a lithotype basis and the gradual changes in these coal properties are presumed to arise primarily from the increase in rank with depth.

The variation of coal properties with lithotype has been examined within continuous sequences of samples taken from five Latrobe Valley coal fields (3). The results indicate that the coal properties related to the organic coal substance, eg: volatile matter, hydrogen, carbon, oxygen and specific energy vary with lithotype layers in the coal seam. The dependence of carbon and hydrogen on lithotype (as measured by color index) is illustrated in Figure 2 for a typical bore. All the major constituents of the organic coal substance are lithotype dependent and their variation within a seam is a direct consequence of the changes in deposit-ional environment which occurred during formation of the seam.

The occurrence of organic sulfur and organic nitrogen is independent of lithotype although the concentration of nitrogen is influenced by the presence of wood in the coal.

Generally speaking, the concentration of minerals in the coal is highest near the overburden and the interseam sediment layers. Because of the discrete nature of the minerals their sporadic distribution in a seam cannot be accurately assessed from a single traverse of sampling through the seam.

The concentration of inorganics, particularly sodium, magnesium, calcium and non-pyritic iron show no relationship to lithotype (Figure 3) indicating that these inorganic species are probably post-depositional in origin. However, some depth related gradients are apparent; for example, sodium and magnesium often show a concentration increase near the top of a bore whilst the aluminium concentration tends to increase near the bottom. This is believed to be due to diffusion of aluminium into the coal from the clay containing sediments below the coal seam. On the other hand calcium in the Morwell open cut has the highest concentration near the middle of the profile.

Variation in Chemical Properties between Coal Fields

In addition to the variation of chemical properties within coal seams, significant variation also occurs between different coal fields in Victoria. An extensive research program in which this variation was investigated has been conducted by the State Electricity Commission of Victoria on behalf of the Victorian Brown Coal Council.

In this Brown Coal Evaluation Programme a sampling philosophy was adopted that would highlight the natural variability of the coal and indicate the range of coal qualities which may be encountered during mining and utilization of the deposits.

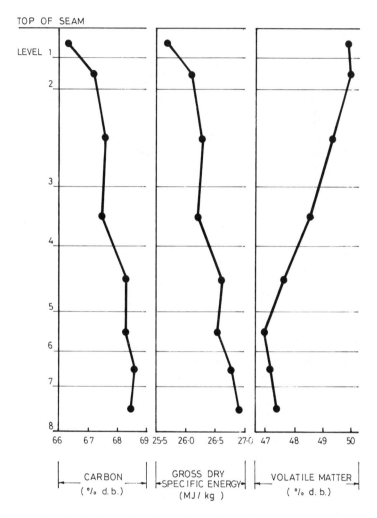

Figure 1. Variation of selected coal properties with depth in the Morwell Open Cut. (Reproduced with permission from Ref. 4. Copyright 1982, Butterworth & Co., Ltd.)

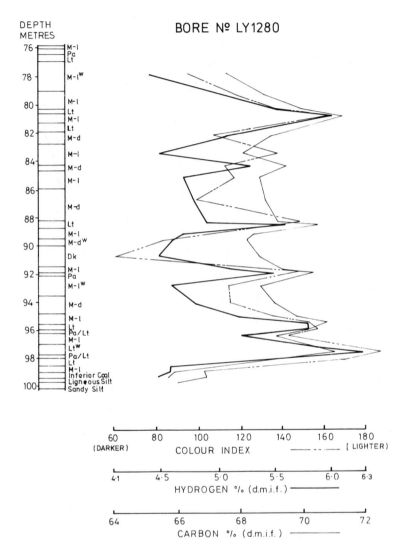

Figure 2. Variation of hydrogen and carbon with depth and color index (lithotype) in a typical bore core.

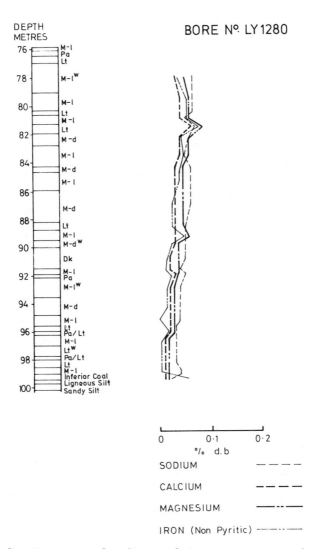

Figure 3. Variation of sodium, calcium, magnesium, and nonpyritic iron with depth in a typical bore core.

The program commenced in the Latrobe Valley coal fields by sampling open-cut faces and 220 mm diameter bore cores and was later extended to include major coal deposits in Victoria. Lithotype logs were prepared using quarter core sections laid out to dry, and selected lithotype samples from each core were analyzed for a variety of chemical, physical and petrographic characteristics and utilization parameters: 144 parameters in all.

To date, a total of 219 coal samples have been analysed from three open-cut faces and twenty 220 mm diameter bore cores, representing 11 coal fields throughout Victoria.   Coal samples from a further six bores are presently being characterized.

The data generated from these analyses and tests was evaluated by statistical methods and correlation coefficients have been determined between all pairs of variables.   Analysis of this data has revealed significant differences in the chemical properties of the coal from different fields and enables the selection of coal with specific properties for particular applications.   Table 4 illustrates some of these differences by showing the range of values determined for selected coal properties from 219 samples (11 coal fields).   Typical values from the two open cut mines currently operating in the Latrobe Valley are shown for comparison.   These differences in chemical properties between the different coal fields are generally more significant than lateral variations within a particular field (ie: between different bores within a field) and they are primarily related to rank.

TABLE 4.   VARIATION   IN   SELECTED   CHEMICAL   PROPERTIES   BETWEEN VICTORIAN BROWN COAL FIELDS

Property	Yallourn Open Cut	Morwell Open Cut	Range of Values (219 Samples)
Moisture (wt %)	66.4	60.2	43.7 - 71.0
Net wet specific energy (MJ Kg^{-1})	6.87	8.95	5.24 - 13.87
Ash (wt %)	1.3	3.5	0.5 - 12.8
Minerals and Inorganics (wt %, db)	1.3	2.7	0.3 - 12.8
Volatile Matter (wt%, dmif)	52.1	49.2	43.3 - 67.8
H/C Atomic Ratio	0.86	0.86	0.77 - 1.16
Oxygen (wt%, dmif)	26.2	24.2	17.4 - 30.0
Nitrogen (wt%, dmif)	0.52	0.62	0.36 - 0.85
Sulphur (wt%, dmif)	0.27	0.34	0.14 - 5.36
SiO$_2$ (wt% db)	0.38	0.14	0.01 - 7.6
Calcium (wt% db)	0.04	0.74	0.01 - 2.07
Magnesium (wt% db)	0.18	0.23	0.02 - 0.85
Sodium (wt% db)	0.06	0.13	0.02 - 0.47
Iron - total (wt% db)	0.18	0.19	0.01 - 1.80
Phenolic - OH (meq/g)	3.72	3.70	1.91 - 4.44
-COOH (meq/g)	2.21	1.90	0.90 - 2.94
-COO$^-$ (meq/g)	0.49	0.59	0.03 - 1.66

dmif - dry, mineral and inorganic-free basis

db - dry basis

Effect of Chemical Properties on Utilization

As a result of the Brown Coal Evaluation Program, the understanding of the variability of Victorian brown coals and its implications for utilization have improved substantially. It has become apparent that certain chemical properties can have important consequences for utilization of the coal for power generation, liquefaction and other applications.
    A number of examples will be briefly described -

Carbonate Formation During Hydrogenation. The formation of carbonate minerals during hydrogenation of low-rank coals can cause serious operational difficulties in the reactor systems. A good correlation has been found between the calcium content of a number of Victorian brown coals and the carbonate formed during hydrogenation. However, the results indicated that cations other than calcium were involved in the formation of carbonate.
    XRD analysis has revealed the presence of several different types of carbonate minerals in liquefaction residues from a number of coals. Minerals identified included vaterite and calcite (two polymorphs of $CaCO_3$) dolomite ($CaMg[CO_3]_2$) and in the residue from a high sulfur coal (2.26% db), anhydrite ($CaSO_4$) was identified. The types of mineral deposits formed depend not only on the coal but also on the reaction conditions. Our data indicates that whilst vaterite forms at low temperatures (380°C), as the temperature increases, the vaterite becomes progressively converted to calcite, the more stable form. After further increases in temperature, particularly at long reaction times, dolomite begins to form.
    The reaction to form anhydrite, in the case of the high sulfur coal, must compete for calcium with the formation of calcium carbonate, and this may have a beneficial effect.
    Clearly the types of inorganic precipitates which form during hydrogenation of Victorian brown coal are dependent on the nature of the exchangeable cations and to some extent the available coal sulfur.

Sodium and Boiler Fouling. The concentration of sodium in coal is regarded as the most significant factor in the formation of troublesome ash deposits during combustion. Although Victorian brown coals are generally low in ash forming constituents, coals with a high proportion of sodium can form ashes which contain large amounts of low melting point, sodium sulphate compounds. These are formed during combustion from the inorganic sodium and organic sulphur in the coal.
    The sodium sulfate condenses on the surface of boiler tubes and together with fly-ash particles forms sticky deposits, which can consolidate on heating and lead to extremely dense hard-to-remove deposits. The presence of high sodium sulfate content ash thus requires special consideration during the design and operation of boilers.

<u>Aluminium and Precipitator Ash.</u>   In some Victorian brown coals significant quantities of acid-soluble aluminium are found.   This is believed to be present as aluminium hydroxide which is dispersed throughout the water phase of the coal.   During combustion of this coal, the refractory aluminium oxide formed takes the shape of the relics of the plant material present in the coal, thus forming an extremely low density ash (approximately 100 kg/m^3).   Whilst the collection of these particles by electrostatic precipitation is possible, the problem of reentrainment on rapping has necessitated the use of larger sized units than would otherwise be required.   It is therefore important to determine the acid soluble aluminium fraction in the coal to determine if precipitation of fly ash is likely to be a problem.

Acknowledgments

The authors wish to acknowledge the support of the State Electricity Commission of Victoria (SECV), the Victorian Brown Coal Council (VBCC), and the companies in the VBCC Industrial Participants Group.   The associated drilling, logging and sampling activities were managed by the SECV's Geological and Exploration Division, and the analytical work was performed by the staff at the Herman Research Laboratory.

LITERATURE CITED

1. Victorian Brown Coal Resource Development Study, Kinhill Pty Ltd.   9 Volumes.   December 1982.

2. Kiss LT and King TN.   <u>Fuel</u> 1977, 56, 340-1.

3. King TN, George AM, Hibbert WD and Kiss LT.   Variation of Coal Properties with Lithotype Pt 2.   SECV Report No. SO/83/55 (1983).

4. Perry, GJ; Allardice, DJ; Kiss, LT <u>Fuel</u> 1982,61,1060.

RECEIVED March 12, 1984

Correlations Between Petrographical Properties, Chemical Structure, and Technological Behavior of Rhenish Brown Coal

E. A. WOLFRUM

Rheinische Braunkohlenwerke AG, Stüttgenweg 2, 5000 Köln 41, Federal Republic of Germany

The brown coal reserves in the Federal Republic of
Germany amount to approx. 56 billion metric tons,
55 billion metric tons of which are in the Rhenish
brown coal district located west of Cologne. About
35 billion metric tons of that reserve are
considered to be technologically and economically
mineable today (1).
Rhenish brown coal is now mined in five opencast
mines with an average depth of about 280 m. Modern
mining equipment having a daily capacity of some
240,000 m³ is used. About 119 million metric tons
of brown coal were mined in 1981.
Of that output, 84.5 % was used for power
generation, 7.9 % for briquette production in 4
briquetting plants of the Rheinische Braunkohlenwerke and 4.2 % for powdered brown coal production
in two grinding mills. A low portion, viz. 0.3 %,
was used in a Salem-Lurgi rotary hearth furnace to
produce about 96,000 metric tons of fine coke in
1981. About 2.7 % of the brown coal output was
used for other purposes, inter alia in test plants
for processes like gasification and hydrogenation.
Mining in greater depths leads to a change of the
geotectonic conditions and hence a natural change
in the brown coal quality characteristics.
This has varying and graduated effects on the
individual refining processes (2).

0097–6156/84/0264–0015$06.00/0
© 1984 American Chemical Society

The petrographical and chemical investigations
presented in the following sections were carried
out in order to describe the behaviour of the coal
types characteristics of the Rhenish brown coal
area during refining processes.

STRUCTURE AND COMPOSITION OF RHENISH BROWN COAL

Rhenish brown coal consists of a variety of
lithotypes which are already discernible in the
coal seam by brightness variations.
Recent pollen analytical investigations proved
that the bright and dark layers result mainly from
changing conversion and decomposition conditions.
The bog facies has only a limited influence.
The gradation from dark to bright layers reflects
the degree of brown coal destruction (3,4).
Only a model can establish the complex,
heterogeneous structure of brown coal.
Figure 1 shows a model that interconnects the
various structural components, namely lignin,
humic acid and aromatic structural elements. The
high content of functional groups causes high
reactivity. Figure 2 shows how oxygen-containing
functional groups are distributed in the Rhenish
brown coal.

Macropetrographical characterization

Based on macropetrographical criteria (determination of the lithotypes by visual examination
with the naked eye) 15 brown coal lithotypes were
selected for the investigations described in the
following; they represent more than 90 % of the
main seam.
Figure 3 shows these 15 brown coal lithotypes
arranged according to stratification (unbanded to
banded coal) and texture (plant tissue content).

Micropetrographical characterization

Micropetrography evaluates the coal components
ascertainable by microscopy. Figure 4 shows an
extract of the results obtained from the combined
maceral-microlithotype analysis after the International Handbook of Coal Petrography of the 15
brown coal lithotypes.
Based on the qualitative assignment of the
individual coal components, Rhenish brown coal can
be divided into four groups having similar

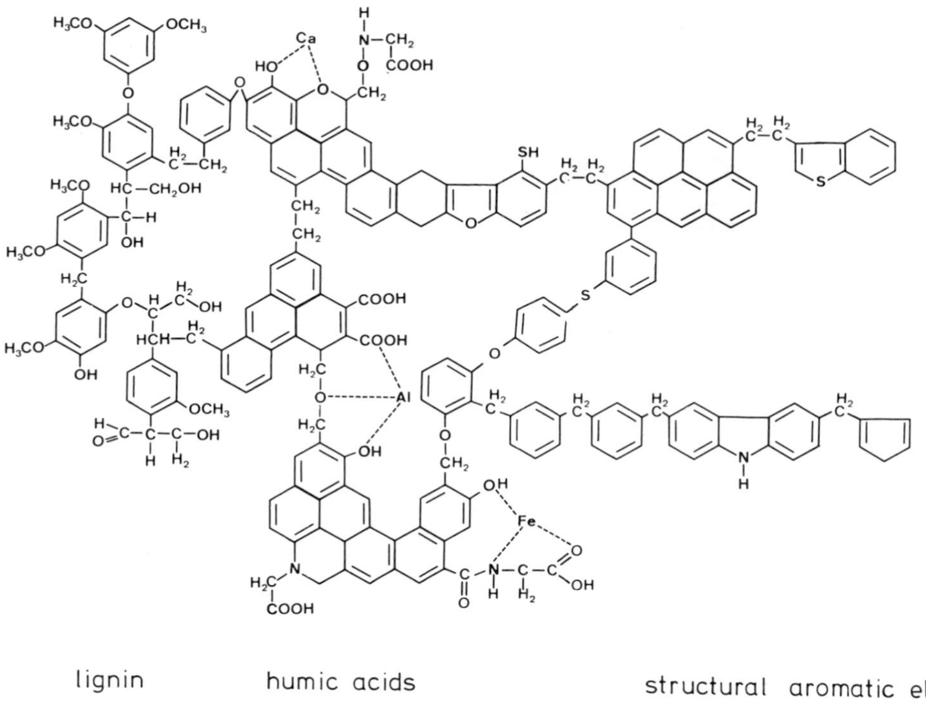

element	Rhenish brown coal (maf)
C	68,7
H	5,05
O	25,0
N	1,0
S (org.)	0,25

typical composition

structural aromatic elements

humic acids

lignin

Figure 1. Schematic composition of the brown coal structure.

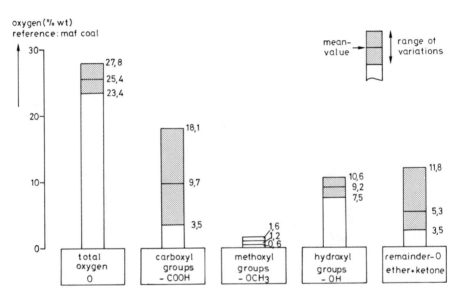

Figure 2. Distribution of the oxygen-containing functional groups in Rhenish brown coal.

percentage of a seam section	sample no.	structure matrix / texture	brightness and colour	gelification rate accessory intercalations
4,2 %	1	unstratified coal	bright	xylite-and mineral-free, matrix
3,9 %	2	unstratified coal	dark	granular gelification nests, xylite
5,5 %	3	unstratified coal	medium-bright	xylite-containing
15,0 %	4	unstratified coal	medium-bright	slightly textured and xylite
12,8 %	5	unstratified coal	medium-dark	slightly textured,xylite,surface gel
7,8 %	6	unstratified coal	medium-bright	slightly textured,xylite-free
5,8 %	7	unstratified coal	dark	slightly textured,xylite,gel,fusite
18,5 %	8	slightly stratified coal	medium-bright	textured,xylite and resinous substance
9,2 %	9	slightly stratified coal	medium-dark	textured,xylite and bright texture
3,0 %	10	slightly stratified coal	dark	xylite and gelled,textured
5,7 %	11	slightly stratified coal	dark	textured,xylite and fusite nests
1,9 %	12	heavily stratified coal	medium-bright	heavily textured,bright texture,resinous sub.
2,5 %	13	heavily stratified coal	dark	heavily textured,xylite and fusite nests
2,2 %	14	heavily stratified coal	dark	heavily textured,xylite,gelled
2,0 %	15	fibrous coal	dark (black)	heavily textured,fusite texture

Figure 3. Macropetrographical classification of 15 brown coal lithotypes from the Rhenish area. (Reproduced with permission from Ref. 2. Copyright 1981, Schriftleitung Braunkohle.)

maceral groups, maceral subgroups, microlithotypes	group no.	1	2	3	4
	lith. no.	1,2	3,4,6,7	5,8,10	9,11,12, 13,14,15
maceral groups					
liptinite	% vol.	26 - 28	20 - 26	6 - 11	3 - 12
huminite	% vol.	67 - 69	72 - 79	87 - 92	80 - 90
inertinite	% vol.	5	1 - 3	1 - 4	2 - 17
maceral subgroups					
humodetrinite	% vol.	56 · 59	54 - 58	31 - 43	12 - 31
humotelinite	% vol.	3 - 5	10 - 17	26 - 33	34 - 50
humocollinite	% vol.	6 - 7	3 - 8	19 - 26	17 - 31
microlithotypes					
lipto-humodetrite	% vol.	79 - 80	56 - 67	3 - 17	-
telo-humodetrite	% vol.	3	21 - 30	8 - 15	10 - 30
telo-humocollite	% vol.	-	-	9 - 15	3 - 17
detro-humocollite	% vol.	-	-	3 - 7	3
gelo-humotelite	% vol.	3	3	10 - 12	10 - 39

Figure 4. Micropetrographical classification of 15 lithotypes into 4 groups. (Reproduced with permission from Ref. 2. Copyright 1981, Schriftleitung Braunkohle.)

micropetrographical properties which correspond
with the macropetrographical features (principle
of classification: stratification and texture).
The classification into four main groups results
in the correlations shown in the following
Figures. Figure 5 shows the liptinite content as a
function of the group classifications 1 to 4.
With rising group number, corresponding to
stronger stratification and texture, the liptinite
content declines. Figure 6, which shows the
huminite as a function of the group numbers,
indicates that the huminite content rises with
increasing stratification and texture.

   Chemical and physical composition

Subsequent to the petrographical coal analysis,
both a chemical and a chemo-physical investigation
were carried out. Figure 7 shows the chemical and
physical properties of the investigated brown coal
lithotypes.
Rhenish brown coal has an average ash content of
about 4 % (mf), a volatile matter of about 52 %
(maf) and a lower heating value of 25.6 MJ/Kg of
coal (maf). The final analysis of the coal under
maf conditions shows the following average
composition:
69 % carbon, 5 % hydrogen, 1 % nitrogen and
approx. 25 % oxygen; the sulphur content amounts
to about 0.35 % (maf), about 50 to 70 % of which
is bound to the ash during the combustion process
since approximately half of the mineral components
of Rhenish brown coal consist of basic alkali and
alkaline earth compounds -primarily those of
calcium.

CORRELATIONS BETWEEN PETROGRAPHICAL STRUCTURE,
CHEMICAL COMPOSITION AND REFINING BEHAVIOUR OF
RHENISH BROWN COAL

The correlations between the chemical brown coal
data, petrographical parameters and the refining
behaviour are described in the following
discussion.
A direct correlation among the quality of the raw
material, the refining cost, and the quality of
the desired products is applicable to nearly all
brown coal refining processes.
Hence, it is indispensible for any refining
operation to know and assess the composition of
the raw material and its behaviour during the
refining process.

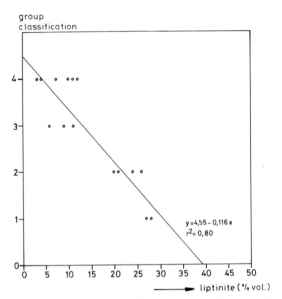

Figure 5. Liptinite content as a function of micropetrographical classification. (Reproduced with permission from Ref. 2. Copyright 1981, Schriftleitung Braunkohle.)

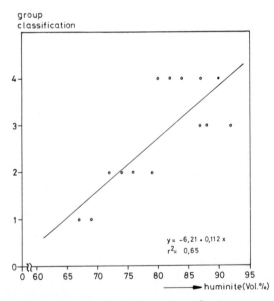

Figure 6. Huminite content as a function of micropetrographical classification. (Reproduced with permission from Ref. 2. Copyright 1981, Schriftleitung Braunkohle.)

The usability of the results obtained from the raw
material characterization in everyday practice
depends on
- the spacing of drillings usable for quality
  assessment
- the level of geological knowledge of the coal
  forming conditions
- a representative sampling in the opencast mine,
  taking into account the cutting geometry of the
  excavator (position of the excavator cuts to the
  deposit), the rate of mining advance and the
  high volumes involved.

Correlations between chemical and petro-graphical parameters

Heating value

Figure 8 shows that the heating value declines
with increasing coal stratification. A comparison
of the heating value of coal and its hydrogen
content in Figure 9 indicates that according to
expectation the heating value of the coal rises
parallel to an increase in its hydrogen content.
This again is due to the portion of hydrogen-rich
macerals, such as liptinite, declining in the said
order - a correlation clearly evident in
Figure 10. The heating value of the coal obviously
rises in proportion to its liptinite content. The
oxygen-rich lignitic coal components, such as
huminite, have the opposite effect on the heating
value: Figure 11 shows that the heating value of
the coal is down sharply with an increase in the
portion of this maceral group.

Volatile content

The individual coal constituents contribute
different portions to the volatile matter of the
coals (5). With increased stratification, the
volatile matter decreases, caused by the relative
distribution of the various maceral groups.
Figure 12 gives an example how the content of
volatile matter and that of the hydrogen-rich
liptinite maceral correlate.

Hydrogen content

Figure 13 shows that also the hydrogen content
closely corresponds to the petrographical properties. Figure 14 shows the correlation between

lithotype no.			1	2	3	4	5	6	7	8	9	10	11	12	13	14	15
Short analysis																	
- brown coal (raw)	an*	%	52,2	60,4	60,1	55,6	52,0	56,6	54,0	60,0	53,4	55,1	54,0	59,7	60,6	58,6	49,0
- ash content	mf	%	6,53	2,44	2,78	3,86	3,90	4,42	3,55	4,27	3,04	3,52	3,73	2,73	3,56	3,38	3,92
- volatiles	maf	%	59,68	55,25	55,11	54,70	52,42	52,09	51,91	52,98	49,65	49,76	48,36	53,72	51,74	51,77	48,79
- C-fix	maf	%	40,32	44,75	44,89	45,30	47,58	47,9¹*	48,09	47,02	50,35	50,24	51,64	46,28	48,26	48,23	51,21
- heating value	maf	MJ/kg	27,99	25,87	25,89	26,69	25,85	25,97	26,30	25,27	25,55	25,43	25,31	24,96	25,43	25,36	24,99
Elementary analysis																	
- carbon	maf	%	69,6	67,5	67,8	68,9	68,2	69,3	69,6	65,7	68,5	68,4	68,9	67,4	66,3	67,4	69,5
- hydrogen	maf	%	5,89	5,60	5,24	5,56	5,19	5,21	5,25	4,71	5,10	4,92	4,69	4,91	4,79	4,88	4,51
Extraction analysis																	
- bitumen content	maf	%	12,87	12,25	10,96	12,04	8,79	7,43	8,96	7,29	7,10	5,40	9,18	7,20	9,65	5,11	7,31
- paraffin content	maf	%	10,19	9,46	9,00	10,00	4,93	6,53	7,48	6,00	5,77	3,93	5,90	5,93	8,20	4,19	3,16

* as received

Figure 7. Chemo-physical characterization of the brown coal lithotypes subjected to investigation. (Reproduced with permission from Ref. 19. Copyright 1983, Schriftleitung Braunkohle.)

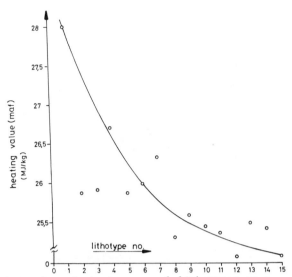

Figure 8. Heating value of the brown coal lithotypes as a
function of macropetrographical classification. (Reproduced
with permission from Ref. 2. Copyright 1981, Schriftleitung Braunkohle.)

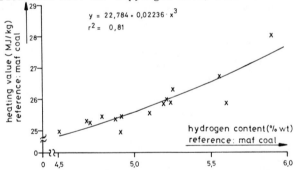

Figure 9. Correlation between the heating values and hydrogen
contents of the 15 lithotypes investigated. (Reproduced with
permission from Ref. 19. Copyright 1983, Schriftleitung Braunkohle.)

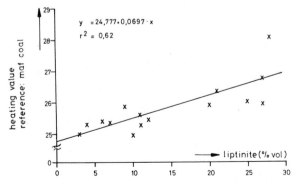

Figure 10. Heating value as a function of the liptinite content
of the 15 lithotypes investigated. (Reproduced with permission
from Ref. 19. Copyright 1983, Schriftleitung Braunkohle.)

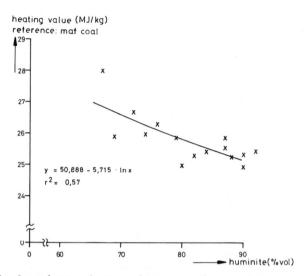

Figure 11. Correlation between heating values and huminite contents of the 15 lithotypes investigated. (Reproduced with permission from Ref. 19. Copyright 1983, Schriftleitung Braunkohle.)

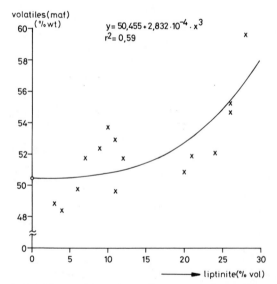

Figure 12. Volatile portion as a function of the liptinite content. (Reproduced with permission from Ref. 19. Copyright 1983, Schriftleitung Braunkohle.)

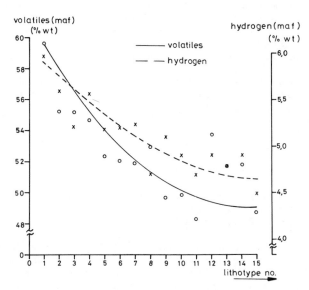

Figure 13. Correlation between volatile matter, hydrogen content
of the lithotypes, and the macropetrographical classification.
(Reproduced with permission from Ref. 2. Copyright 1981,
Schriftleitung Braunkohle.)

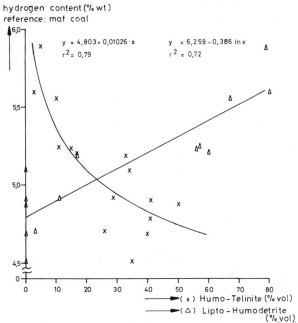

Figure 14. Correlation between hydrogen and Humo–Telinite
and Lipto–Humodetrite portions. (Reproduce with permission
from Ref. 2. Copyright 1981, Schriftleitung Braunkohle.)

the hydrogen content and the lipto-humodetrite
and/or the humotelinite content of the investigated lithotypes. A higher amount of lipto-
humodetrite leads to an increased hydrogen
content. The content of the humotelinite maceral,
however, has a totally different effect: This
component has a high portion of oxygen-rich
molecular groups which causes the hydrogen content
of the brown coal to drop.

Correlations between coal quality and
refining behaviour

Briquetting and coking

There are many investigations and publications
available on the briquetting behaviour of brown
coals, both from the GDR and the Rhenish area (6
to 9).
In order to establish statistically usable data on
the briquetting behaviour of Rhenish brown coals,
the 15 brown coal lithotypes were briquetted under
identical conditions (water content, grain size
distribution and mould pressure) with a laboratory
press.
For assessing the briquettability of these coals,
a number of briquetting parameters were correlated
with the petrographical properties of the brown
coal types. A statistic evaluation of these
briquetting parameters and the micropetrographical
composition of the coals reveals only a minimal
degree of interdependence. Figure 15 clearly shows
that the humotelite content correlates with the
height and volume expansion of the briquettes. The
briquetting expenditure is related to the telo-
humocollite content. The correlation coefficient
(r) varies between 0.76 and 0.82, and is comparatively low. All the other correlations between
the briquetting parameters (e.g. diametrical
expansion) and the petrographical coal composition
turned out to be rather insignificant so that they
need not be taken into consideration.
These investigations on correlations established
between the raw material properties and
briquettability of Rhenish brown coal led to the
following results:
1. A macro- and micropetrographical analysis with
   a view to before technological problems
   involved in the briquetting process allows one
   at the most to judge the briquettability of
   Rhenish brown coal on the basis of trend data.

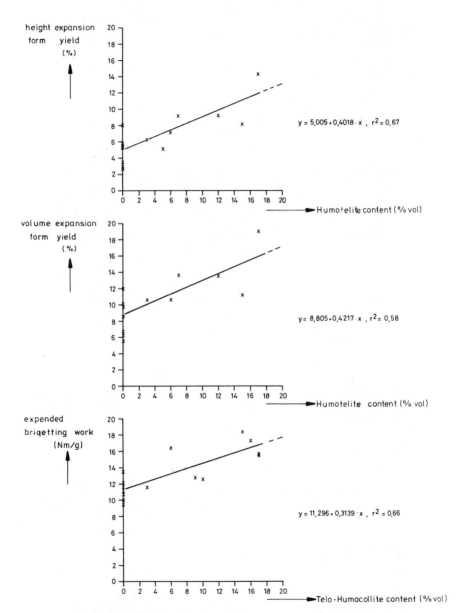

Figure 15. Correlation between the contents of various microlithotypes and various briquetting results. (Reproduced with permission from Ref. 19. Copyright 1983, Schriftleitung Braunkohle.)

2. A correlation analysis of the chemo-physical
   parameters of Rhenish brown coal and its
   briquettability gives only trend data as well.
Therefore generally an anticipated quality
assessment of the briquettes is restricted to the
following points:
1. Macropetrographical assessment of the coal seam
   and evaluation of the briquettability on the
   basis of values gained by experience.
2. Laboratory production of briquettes and
   determination of their compression strength.
3. Determination of the ash content as an essential factor for the assessment of brown coal
   briquettability.
Numerous publications (10 to 12) have appeared,
principally from the GDR, on the required quality
properties of brown coal and their influence on
the quality characteristics of formed coke. Since
the Rheinische Braunkohlenwerke AG is not engaged
in formed coke production at present, raw material
quality and coking behaviour are of interest only
for the production of fine coke using the rotary
hearth furnace (13, 14).
This technology is dependent only to a small
extent on the specific raw material composition.
The petrographical factors of the feedstock have
an impact principally on grain size and grain size
distribution.

Gasification

Two gasification processes under development,
namely High Temperature Winkler Gasification (HTW)
and hydrogasification of lignite (HKV), encouraged
to study the gasification behaviour of various
brown coal lithotypes (15).
HTW process principle: Fine-grained dry brown coal
is gasified in a fluidized bed with oxygen/steam
or air under pressure and at a high temperature
into a gas rich in carbon oxide and hydrogen.
HKV process principle: Fine-grained dry brown coal
is gasified in the fluidized bed with hydrogen
under high pressure and at a higher temperature
into a CH_4-rich gas.
The reactivity of the residual char is the
rate-controlling factor for brown coal hydrogasification (16).
Laboratory-scale investigations on the gasification behaviour of various types of brown coal
coke showed that the mode of pretreatment has a

greater influence on the gasification process than
do the raw material properties. To give an
example, thermal pretreatment of the coke in an
inert atmosphere like helium at gasification
temperature (900 °C) has a very favourable effect
on gasification.
With one exception only, gasification degrees
close to 100 % were achieved.
Differences in gasification rates are due to the
heterogeneous pore structure and the inhomogeneous
iron distribution in the coal matrix.
Irregularly localized iron groupings contained in
cokes of the lithotypes 1 and 8 show a varying
reactivity behaviour.
Cokes of the lithotypes 4, 5, 9, 11, 13 and 15
with similar gasification behaviours have a
comparatively homogeneous distribution of all ash
components (Fig. 16).
No correlation was established between the maceral
composition and the reactivity behaviour of the
cokes (17).
As was the case with the mentioned hydrogasification process, the results obtained with HTW
gasification did not show any statistically
significant correlations between the petrographical composition of the lithotypes and their
gasification behaviour.

Liquefaction

To determine potential raw material impacts on
brown coal hydroliquefaction, the 15 lithotypes
were converted into liquid products using various
techniques (18):
1. moderate indirect hydrogenation with tetralin
   as a hydrogen-transferring solvent
2. direct hydrogenation with hydrogen and different catalysts similar to operational conditions.
Indirect hydrogenation using tetralin at a
reaction temperature of 410 °C, a pressure of 400
bar and an overall reaction time of 2 h produced
carbon conversion rates from 50 to 79 (%wt) and
liquid product yields from 43 to 69 (%wt). Of the
multitude of correlations established between the
results of the hydrogenation tests and the
micropetrographical composition of the coal types
only a few examples are given.
Figure 17 shows the product yields of indirect
brown coal hydrogenation with tetralin as a
function of stratification and texture (lithotype

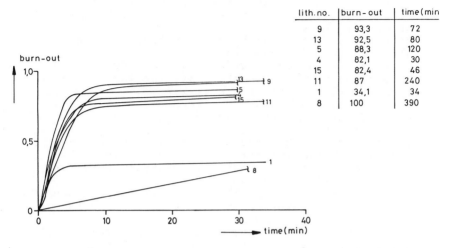

40 bar H_2/900°C

pretreatment: 0,5 h, 1 bar (helium), 900°C

lith. no.	burn-out	time (min)
9	93,3	72
13	92,5	80
5	88,3	120
4	82,1	30
15	82,4	46
11	87	240
1	34,1	34
8	100	390

Figure 16. Conversion as a function of time with hydrogasification of brown coal cokes. (Reproduced with permission from Ref. 19. Copyright 1983 Schriftleitung Braunkohle.)

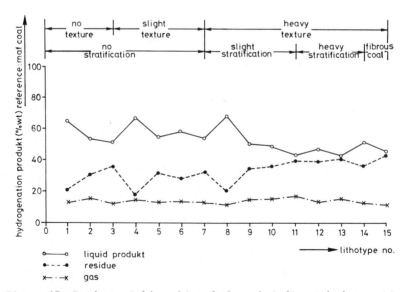

Figure 17. Product yields achieved through indirect hydrogenation with tetralin as a function of stratification and texture (lithotype number). (Reproduced with permission from Ref. 19. Copyright 1983, Schriftleitung Braunkohle.)

number). It is obvious that with increasing
stratification and texture, i.e. with rising
lithotype number, the liquid product yield drops
and the portion of hydrogenation residue
increases. Regarding the fine structure of the
coal the following correlations result:
Figure 18 represents the carbon conversion rate as
a function of the liptinite and huminite maceral
portions. It can be seen that an increase in the
hydrogen-rich liptinite constituent improves
carbon conversion while the oxygen-rich huminite
reduces the carbon conversion rate. These trends
also apply to the yield of liquid products.

During direct hydrogenation with molecular
hydrogen and catalyst, the influence of the raw
material was expected to weaken using hydrogen and
a catalyst under the conditions similar to those
in the real hydroliquefaction process. An increase
in the carbon conversion rate up to a maximum of
96 (%wt) shows that brown coals with high or
low carbon-to-hydrogen ratios achieve high product
yields. Distinct impacts of the raw material on
the liquefaction results are no longer observed.

Summarizing, it can be said that the extent of
hydrogenation rises in direct proportion to carbon
conversion. The result is that nearly all brown
coal types mineable in the Rhenish area can be
converted into liquid products with high yields.
The investigations showed that in general practice
no importance is attributed to a micropetrographical assessment of the coal types as a criterion
for selecting specific brown coals from the
Rhenish area. In general, brown coals from various
areas that meet the quality parameters given in
Figure 19 have satisfactory hydrogenation properties.

SUMMARY

For the purpose of an assessment with a view to
refining, the petrographical, chemical and
physical properties of lithotypes of Rhenish brown
coal were established and compared with one
another.
The investigated coal types cover more than 90 %
of the coal types determined by boring in the
Rhenish deposit. A correlation of the results
shows a describable, sometimes multidimensional,
dependancy.

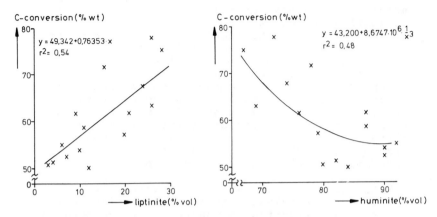

Figure 18. Carbon conversion achieved through indirect hydrogenation with tetralin as a function of the liptinite and
huminite contents. (Reproduced with permission from Ref. 19.
Copyright 1983, Schriftleitung Braunkohle.)

1. micropetrographical properties

liptinite content	> 20 % vol
lipto- humodetrite content	> 50 % vol
humodetrinite content	> 50 % vol
inertinite content	< 3 % vol
gelo-humotelite content	< 6 % vol
diffuse reflectance	> 7 %

2. chemical properties

tar content	> 11 % wt (mf)
paraffin content	> 7 % wt (maf)
hydrogen content	> 5 % wt (maf)
C/H ratio	< 1,1
volatiles	> 53 % wt (maf)
carbonization water	< 8,7 % wt (mf)

Under the conditions stated above, the 5th and 6th
seam of the Fortuna-Garsdorf mine contain approx.
78 % wt of good hydrogenation coals

Figure 19. Petrographical and chemical properties of good
hydrogenation coals. (Reproduced with permission from Ref. 19.
Copyright 1983, Schriftleitung Braunkohle.)

A comparison of raw material properties and the
results of the technical experiments quickly
reveals the limits set to the petrographic
approach.

Out of all the refining processes subjected to
investigation briquetting places the highest
requirements on the raw material properties.
For the gasification process no usable quantitative relations were established between the
petrographical coal properties and the gasification behaviour. It is possible without any
material problems to convert nearly all the
Rhenish brown coal types into gaseous hydrocarbons
or synthesis gas.
This statement applies without any restriction to
coal liquefaction as well. Hence, for hydrogenation process engineering the optimization of the
reaction conditions has precedence over the raw
material properties.
Under the given conditions of the Rhenish brown
coal deposit, an opencast mining operation, a high
output and the large feed quantities required for
future refining plants brown coal petrography is
one out of many tesserae for quality assessment.
The development of appropriate modes of determining the quality characteristics of raw brown
coal is a task indispensable for the future.

Acknowledgments

We thank H. W. Hagemann, PhD., member of the
faculty of geology, geochemistry and oil and coal
deposits of the Aachen technical university for
having carried out the micropetrographical
investigations.

APPENDIX

(1971, 1975): International handbook of coal
petrography -Intern. Committee for Coal Petrology
(ed.), Suppl. to 2nd., Centre National Recherche
Scientifique: Paris.

Literature Cited

1) P. Speich:
"Technische und wirtschaftliche Gesichtspunkte der
Braunkohlenveredlung"
Brennstoff-Wärme-Kraft 32 (1980) Nr. 8, 307-312

2) E. Wolfrum, J. Wawrzinek:
"Beziehungen zwischen petrographischen und
chemischen Eigenschaften Rheinischer Braunkohle"
Braunkohle Heft 11/81, 381-386
3) G. von der Brelie, M. Wolf:
"Zur Petrographie und Palynologie heller und
dunkler Schichten im rheinischen Hauptbraunkohlen-
flöz"
- Fortschr. Geol. Rheinld. u. Westf. (1981) 29;
95-163
4) H. W. Hagemann, A. Hollerbach:
"Relationship between the macropetrographic and
organic geochemical composition of lignites"
- Advances in organic Geochemistry 1979,
A. G. Douglas, J. R. Maxwell (Editors)
Pergamon Press Oxford, New York, Frankfurt
5) R. Kurtz, E. Wolfrum, W. Assmann:
"Die Verkokungseigenschaften chemischer Stoff-
gruppen in einigen Braunkohlentypen des rheini-
schen Reviers"
Comm. Eur. Communities, Rep EUR 1977
EUR 6075, Round Table Meet. Chem. Phys.
Valorisation Coal, 125-150
6) E. Sonntag, M. Süß:
"Beispiel petrologischer Untersuchungen zur
Klärung rohstoffabhängiger verfahrenstechnischer
Probleme der Braunkohlenveredlung"
Bergbautechnik 19 (1969) S. 255-260
7) R. Kurtz:
"Zusammenhang zwischen mikropetrographischem
Aufbau der Weichbraunkohlen und ihrer Brikettier-
barkeit"
Dissertation, RWTH Aachen, Fakultät für Bergbau
und Hüttenwesen (1970)
8) E. Rammler:
"Zur Kenntnis der Brikettierbarkeit von Braun-
kohle"
Freiberger Forschungshefte A99 (1958), 81-104
9) H. Jakob:
"Neuere Ergebnisse der angewandten Braunkohlen-
petrologie"
Berg- und Hüttenmännische Monatshefte 105, Heft 2
(1960), 21-29.
10) E. Rammler, G. Bilkenroth:
"Zur technischen Entwicklung der Braunkohlen-
kokerei"
Freiberger Forschungshefte A 184 (1961), 85-113
11) E. Rammler, W. Fischer, G. Hänseroth:
"Schrumpfverhalten von Braunkohlen Feinstkorn-
briketts bei der Verkokung bis 1 000°C"
Freiberger Forschungshefte A453 (1968), 121-145

12) W. März:
"Untersuchungen zum Verlauf der Verkokung von Braunkohlenbriketts, insbesondere bei der Spül-gasverkokung"
Freiberger Forschungshefte A375 (1966), 79-99

13) R. Kurtz:
"Koksherstellung aus Braunkohle"
Braunkohle Heft 11, Nov. 1975, 352-356

14) E. Scherrer:
"Herstellung von Braunkohlenkoks im Salem-Lurgi-Herdofen"
Braunkohle Heft 7, Juli 1981, 242-246

15) H. Teggers, K.A. Theis, L. Schrader:
"Synthesegase und synthetisches Erdgas aus Braunkohle"
Erdöl und Kohle-Erdgas ver. mit Brennstoff-Chemie Bd. 35, Heft 4, April 1982, 178-184

16) L. Schrader, H. Teggers, K.A. Theis:
"Hydrierende Vergasung von Kohle"
Chem.-Ing.-Tech. 52 (1980) Nr. 10, 794-802

17) M. Ghodsi, J.P. Lempereur, E. Wolfrum:
"Etude cinetique de l´hydrogenation directe des cokes de differents lithotypes de lignite"
Proceedings of the International Conference on Coal Science,
Düsseldorf, Germany, September 7-9, 1981, 191

18) E. Wolfrum, J. Wawrzinek, W. Dolkemeyer:
"Einfluß der Rohstoffeigenschaften auf die Verflüssigung Rheinischer Braunkohlen"
Proceedings of the International Conference on Coal Science,
Düsseldorf, Germany, September 7-9, 1981, 434

19) Wolfrum, E. A. Braunkohle Heft 1983, June, 182-88.

RECEIVED June 21, 1984

The Structure and Reactions of Northern Great Plains Lignites

W. R. KUBE, HAROLD H. SCHOBERT, S. A. BENSON, and F. R. KARNER

Energy Research Center, University of North Dakota, Grand Forks, ND 58202

Extensive deposits of lignite, amounting to over 3.5 x 10^{11} metric tons, occur in the Northern Great Plains of the United States. These coals typically contain over 35% moisture and have a heating value of about 16 MJ/kg. Characterization studies of the organic structure have relied on differential scanning calorimetry as the principal technique. The aromatic clusters generally contain no more than 1 or 2 aromatic rings. Oxygen-containing functional groups are abundant; these coals typically contain 20% oxygen (dmmf basis). The principal groups are carboxylic and phenolic. The inorganic constituents occur in at least three modes and have been studied primarily by chemical fractionation and scanning electron microscopy. The alkali and alkaline earth elements are present as ion-exchangeable cations associated with carboxylic groups. The principal extraneous minerals are quartz, clays, and pyrite. The unique features of both the organic structure and distribution of inorganic species have significant effects on the chemistry of combustion, gasification, and liquefaction.

The lignites of the Northern Great Plains are in the Fort Union Region, which contains the largest reserves of lignite of any coal basin in the world. The Fort Union Region encompasses areas of North Dakota, South Dakota, Montana, and Saskatchewan. The identified resources of lignite in this region amount to 422 Gt (465 billion short tons), of which 24 Gt (26 billion short tons) constitute the demonstrated reserve base (1).

Until about 1970, the utilization of low-rank coals (lignite and subbituminous) was limited, accounting for no more than 1%-2% of the total annual U.S. coal production. In recent years, their production has increased dramatically, and by 1980 production accounted for about 24% of the total national coal production. It has been estimated that in another ten years, low-rank coals could amount to half the total coal production (1).

0097–6156/84/0264–0039$06.00/0
© 1984 American Chemical Society

Lignites exhibit properties which can have profound effects on utilization. These include a high moisture content, high quantities of oxygen functional groups in the carbon structure, an alkaline ash, and inorganic cations attached to carboxylic acid groups. The rapid expansion of lignite utilization in recent years has brought with it an increasing realization of the importance of developing a better understanding of the organic and inorganic structures in lignite and of how these structural features influence lignite reactivity or processing behavior. Results from some current studies in progress in our laboratories on lignite structure and reactivity are presented.

Comparison of Northern Great Plains Lignites with Bituminous Coals

The predominant position of bituminous coal in the total U.S. coal production has resulted, not unreasonably, in the properties of bituminous coals being more extensively studied and thus better known to the general coal research community than those of lignites.

The average proximate and ultimate analysis of Fort Union lignites are summarized in Table I, together with average values for a Pittsburgh seam bituminous coal ($\underline{2}$, $\underline{3}$). The important points to note are the much higher moisture content, higher oxygen, and lower heating value of the lignite.

Table I. Average Analyses of Northern Great Plains Lignite and Bituminous Coal Samples

| | Northern Great Plains Lignite | | Pittsburgh Seam Bituminous | |
	As-Received	Moisture-Ash Free	As-Received	Moisture-Ash Free
Proximate, %				
Moisture	37.2	--	2.3	--
Volatile Matter	26.3	46.5	36.5	39.5
Fixed Carbon	30.3	53.5	56.0	60.5
Ash	6.2	--	5.2	--
Ultimate, %				
Hydrogen	6.9	4.9	5.5	5.6
Carbon	40.7	71.9	78.4	84.8
Nitrogen	0.6	1.1	1.6	1.7
Oxygen	45.0	21.0	8.5	7.0
Sulfur	0.6	1.1	0.8	0.9
Heating Value, MJ/kg	15.9	28.1	32.6	35.2
Reference Source:	($\underline{2}$)		($\underline{3}$)	

Usually, lignitic ash contains a much higher proportion of alkali and alkaline earth elements, and consequently lower proportions of acidic oxides such as silica and alumina, than does ash from bituminous coals. These differences are illustrated by the data in Table II (1). The data on spruce bark ash are taken from reference (4) and show the similarity of lignitic and woody ashes. Of interest are the contrasts between the lignite and bituminous averages first with regard to silica and alumina and, second, to lime, magnesia, and sodium oxide.

Table II. Average Ash Compositions of Northern Great Plains Lignites and Bituminous Coals, SO_3-Free Basis

	Spruce Bark	Lignite	Bituminous
Acidic Components:			
SiO_2	32.0	24.4	48.1
Al_2O_3	11.0	13.8	24.9
Fe_2O_3	6.4	11.3	14.9
TiO_2	0.8	0.5	1.1
P_2O_5	--	0.4	0.0
Basic Components:			
CaO	25.3	30.5	6.6
MgO	4.1	8.6	1.7
Na_2O	8.0	8.1	1.2
K_2O	2.4	0.5	1.5
Reference Source:	(4)	(1, 2)	(1)

Organic Structural Relationships

Lignite is an early stage in the coalification process and thus could be expected to retain some characteristics of wood. This relationship is illustrated by the electron micrographs shown as Figures 1 and 2. Pieces of plant debris, presumably twigs or rootlets, can be seen in Figure 1. Remains of the woody cellular structure are visible in Figure 2. This structure is similar to that of a softwood (5).

On the molecular level, the distinguishing features of the organic structure of lignite are the lower aromaticity (or fraction of total carbon in aromatic structures) compared to bituminous coals; aromatic clusters containing only one or two rings; and the prevalence of oxygen-containing functional groups. A proposed structural representation, modified slightly from the original version (6), is given in Figure 3. We do not claim that this actually represents the structure of Northern Great Plains lignite, but does illustrate the major structural features.

The aromaticity has been studied by pressure differential scanning calorimetry (PDSC). The details of the experimental technique and of the methods for calculating aromaticity from a PDSC thermogram

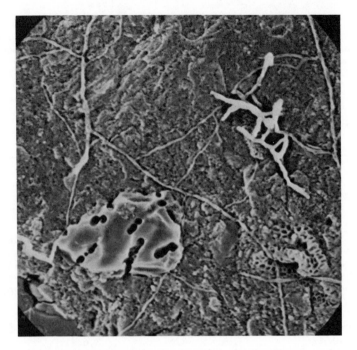

Figure 1. Electron micrograph of ion-etched Beulah lignite,
showing rootlets or other plant debris (5850X).

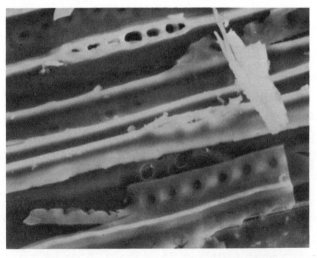

Figure 2. Electron micrograph of woody lithotype of Beulah
lignite showing cellular structure (290X). Compare softwood
structure in Reference 5.

have been published elsewhere (7). Briefly, the PDSC experiment provides for controlled combustion of a 1-1.5 mg sample of -100 mesh coal in a 3.5 MPa atmosphere of oxygen. The sample is heated at 20°C/min in the range of 150° to 600°C. The instrument response is a thermogram plotting heat flux, Δq, versus temperature, the integrated value thus being the heat of combustion. For most coals, and many organic compounds and polymers, the thermogram in this region consists of two peaks, which, from comparison to the behavior of model compounds, arise primarily from combustion of the aliphatic and aromatic portions of the sample. The aromaticity may be deduced from a comparison of peak heights.

The aromaticity for several samples of Northern Great Plains lignites, as measured on run-of-mine material, lies in the range of 0.61 to 0.66. A sample of Australian brown coal had an aromaticity of 0.56; a sample of Minnesota peat had an aromaticity of 0.50. Samples of vitrinite concentrates from the Northern Great Plains lignites were more aromatic, with measured values from 0.72 to 0.74.

The temperature at which the maximum of the aromatic peak occurs has been shown to be a function of the extent of ring condensation (7), with the maximum shifting to higher temperatures with increasing condensation. In support of the studies on coal structure, we have measured the PDSC behavior of over 30 organic compounds (most of which have been suggested as coal models or have been identified in the products of coal processing) and about 50 polymers. The maxima of the aromatic peaks in the thermograms of Northern Great Plains lignites generally fall into the same temperature range (375°-400°C) as those for compounds or polymers having benzene or naphthalene rings. One example is given in Figure 4, in which the PDSC thermograms of Gascoyne (N.D.) lignite and poly(4-methoxystyrene) are compared. We conclude that the aromatic structures therefore are mostly one- or two-ring systems.

Much less is known about the hydroaromatic structures or aliphatic bridges between ring systems. A methylene bridge is often suggested as a typical aliphatic bridging group, and was originally shown in the proposed structural representation (6). However, considerations based on thermochemical kinetics predict a half-life of 10^6 years for bond cleavage of diphenylmethane in tetralin at 400°C (8). Exhaustive analyses of the products from liquefaction of Northern Great Plains lignite at 400°C and higher in the presence of tetralin (see (9) for example) have never identified diphenylmethane or related compounds. Absence of diphenylmethane constitutes strong circumstantial evidence for the relative unimportance of methylene linkages between aromatic clusters.

Oxygen is distributed among carboxylate, phenol, and ether functional groups. The carboxylate concentration has been measured by reaction of demineralized coal with barium acetate following the procedure of Schafer (10). Data for seven Northern Great Plains lignite samples indicate carboxylate concentrations ranging from 1.46 to 3.58 meq/g on a dry, mineral-matter free basis. Concentrations of phenolic or ether functional groups have not yet been measured. Electron spectroscopy for chemical analysis (ESCA) provides a means for discriminating between carbon atoms incorporated in C=O and C-O structures by a Gaussian-Lorentzian decomposition of the carbon 1s spectrum (11). Figure 5 provides a comparison of the decomposed carbon 1s spectra of Beulah (N.D.) lignite and polyethylene tereph-

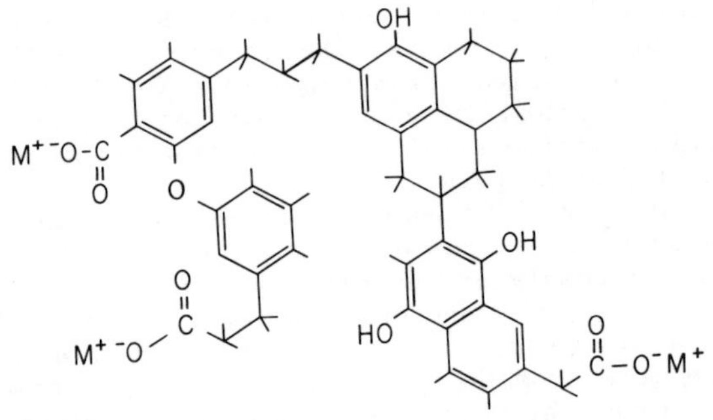

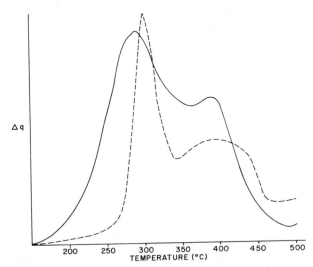

Figure 3. Proposed representation of structural features of lignite. (Adapted from Reference 6.)

Figure 4. PDSC thermograms of Gascoyne lignite (——) and poly(4-methoxystyrene) (--). Plot is of heat flux in arbitrary units vs. temperature (C).

thalate. At present, the ESCA data cannot be resolved into phenolic and etheric carbons. However, it can be shown that the ratio of carbon in carboxylate groups to that in (phenol plus ether) groups is about 0.62 (11). The presence of methoxy groups has been confirmed; as-yet unpublished work by E.S. Olson and J.W. Diehl demonstrates the production of methanol from sodium periodate and peroxytrifluoro-acetic acid oxidations of lignite.

Little consideration has yet been given to the three-dimensional structure. A preliminary examination of lithotypes of Beulah lignite has been conducted by laser Raman spectroscopy (11). The lithotype having a higher concentration of carboxylic acid groups has a weaker band at 1600 cm^{-1}. If this band is assigned as a graphite mode (12), results suggest that the relatively bulky carboxylate groups, with their associated counterions, may disrupt, or preclude, three-dimensional ordering.

Distribution of Inorganic Constituents

In lignites the inorganic constituents are incorporated not only as discrete mineral phases, but also as relatively mobile ions, presumably associated with the carboxylic acid functional groups. The distribution of inorganic constituents has been studied principally by the chemical fractionation procedure developed by Miller and Given (13).

Extraction of the coal with 1M ammonium acetate removes those elements present on ion exchange sites, which are presumed to be carboxylic acid functional groups. Sodium and magnesium are incorporated almost exclusively as ion-exchangeable cations. For a suite of ten Northern Great Plains lignites tested, 84% to 100% of the sodium originally in the coal and 88% to 90% of the magnesium are removed by ammonium acetate extraction. Figure 6 is an electron micrograph showing an electron backscatter image due to the presence of sodium intimately associated with the organic material. Calcium is largely present in cationic form, 48% to 76% being extracted. Some potassium is also extracted in this step, in amounts ranging from 20% to 57%.

Further treatment of the ammonium acetate extracted coal with 1M hydrochloric acid removes elements present as acid-soluble minerals or possibly as acid-decomposable coordination compounds. This acid extraction removes essentially all of the calcium and magnesium not removed by ammonium acetate. This finding is suggestive of the presence of calcite or dolomite minerals, which are known to be present in Northern Great Plains lignites (14). The hydrochloric acid extraction behavior of other major metallic elements is quite variable, which suggests significant differences in the mineralogy of the samples. Of those elements not extracted at all by ammonium acetate, some iron, aluminum, and titanium are removed by hydrochloric acid.

The portions of elements which are not removed by either reagent are considered to be incorporated in acid-insoluble minerals, particularly clays, pyrite, and quartz. This group includes all of the silicon, the remaining sodium and potassium, and the residual iron, aluminium, and titanium. The acid-insoluble minerals are present as discrete phases. Frequently the mineral particles are quite small (see Figure 7, for example) and very highly dispersed through the

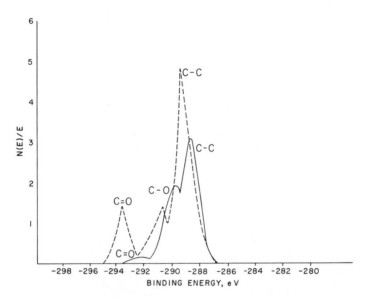

Figure 5. Decomposed ESCA carbon is spectra of Beulah lignite
(—) and polyethylene terephthalate (--). Plot is of a number
of electrons per energy in arbitrary units vs. binding energy
in electron volts. The lignite spectrum has been corrected
for sample charging (11).

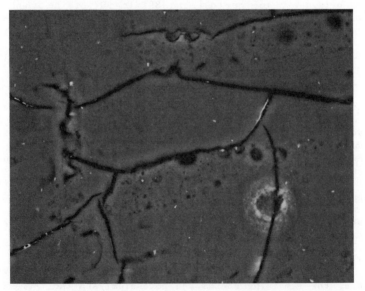

Figure 6. Electron micrograph of Beulah lignite showing
organic region enriched in sodium (circular structure in lower
center of view).

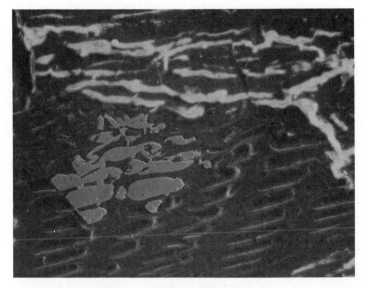

Figure 7.   Electron micrograph of pyrite particles intergrown in carbonaceous structure of Beulah lignite, suggesting difficulty of removal by float/sink (150X).

carbonaceous materials, to such an extent that only about 15%-30% of
the discrete mineral matter is separable in a traditional float/sink
experiment.

Effects of Structure on Reactivity

The small aromatic clusters, the high concentration of organic oxygen
functional groups, and the presence of inorganic species as ion-
exchangeable cations are unique features of low-rank coals. Each of
these features should influence the reactivity of low-rank coals,
thereby giving low-rank coals distinctly different reactions when
compared to bituminous coals. The reactivity indeed has unique
features. As yet, however, little has been done in a deliberate way
to develop an understanding of the connections between structure and
reactivity.

The carboxylic acid group is one of the most important of the
organic functional groups. The carboxylic acid group is labile under
relatively mild thermal conditions, decomposing to carbon dioxide
below 450°C (15). Although the "volatile matter" content, as measured
by the standard proximate analysis, is higher in lignite than in
bituminous coals, much of the lost volatile matter is actually carbon
dioxide rather than hydrocarbon gases, oil, or tar. Only about 17%
of the calorific yield occurs in volatile products from North Dakota
lignite, whereas about 30% occurs in the volatiles from bituminous
coals (16). The ability of the carboxylic acid groups to bind
cations on ion exchange sites has already been mentioned.

An indication of the importance of oxygen in differentiating the
chemistry of low-rank from bituminous coal is provided by electron
spin resonance spectroscopy (ESR) (17, 18). The ESR experiment
provides a measure of the so-called g-factor, which is a dimension-
less constant indicative of the chemical environment of the free
radicals in the sample. The g-factors of bituminous coals are near
the expected values for radicals in highly aromatic systems. The
g-factors of low-rank coals are higher and are indicative of participation by oxygen (or sulfur) in free radical formation. The
participation of oxygen functional groups in radical formation may
influence the chemical reactions of the coal. Such radicals have an
enhanced ability to form adducts with solvent molecules, hence as-
sisting the solvation of the coal. The abstraction of hydrogen from
potential hydrogen-donor molecules in the liquid phase can be facilitated; quinones, for example, are considered to be effective in
hydrogen abstraction.

The facile thermal decarboxylation of carboxylic acid functional
groups can remove much of the oxygen from the lignite without re-
quiring any net consumption of a reducing agent. Thus during liquefaction low-rank coals could actually consume less hydrogen than
bituminous coals, despite having a higher oxygen content. (Of
course, on a carbon-equivalent basis, the low-rank coals would pay
the penalty of having a higher consumption of coal.)

Oxygen functional groups can promote β-bond scission (8), which
may be an important process in the degradation of the coal structure.
The role of ether, carboxyl, and other groups in wood pyrolysis and
combustion has been discussed (19); it seems reasonable to assume
that analogous reactions would occur in low-rank coals. Other possible roles for oxygen functional groups include ether cleavage, and

the cleavage of aliphatic bridges linked to aromatic rings bearing a phenolic group.

The inorganic constituents of low-rank coals have very significant impacts on utilization and conversion processes, so much so that a consideration of their properties and behavior is at least as important as for the carbonaceous portion of the coal. Currently all low-rank coal production is consumed in combustion processes. The most dramatic, and notorious, effect of the inorganic matter in combustion is the fouling of boiler tubes.

Ash fouling, the accumulation of deposits on boiler tube surfaces in utility boilers, is a severe operating problem in many power plants fired with low-rank coals. Ash fouling is a complex phenomenon the extent of which is related to the boiler design, the method of operating the boiler, and the coal properties. In extreme cases it is necessary to schedule frequent shutdowns for removing the deposits or to derate the boiler. A recent survey of six power plants estimated that total costs of curtailments due to ash-related problems were $20.6 million over a six-month period (20).

Ash fouling appears to be initiated by the formation of a layer of sodium sulfate on the boiler tube. It is thought that thermal decomposition of sodium salts of carboxylic functional groups in the coal is the start of a sequence of reactions leading ultimately to the formation of sodium sulfate in the flame or flue gas. The convective mass transfer diffusion of the sodium-containing species through a boundary layer around the tube results in deposition of sodium sulfate on the tube surface.

The deposition of the sodium sulfate layer is followed by accumulation of a layer of ash particles which probably builds up via inertial impaction. This layer of particles gradually thickens until a point is reached at which heat loss to the tube is sufficiently slow to allow crystallization reactions or reactions with gas phase sodium-containing species to occur. At that point a glassy-appearing matrix begins to form. This matrix material, and ash particles trapped in it, constitutes the bulk of the ash deposit.

The matrix consists of complex silicates, of which sodium melilite is an important constituent. Sodium melilite is a low-melting material which could stay fluid long enough to react with other silicates or sulfates and which could crystallize into a hard, strong deposit. Roughly, then, the more sodium in the coal, the more melilite which could form; the more melilite, the more matrix; the more matrix, the worse the fouling problem. Thus there is a rule-of-thumb correlation between the amount of sodium in the coal and the severity of the fouling. Generally coals having less than 2% Na_2O in the ash are low fouling; 2%-6%, medium fouling; and above 6%, high fouling. Above about 10% there is a "saturation effect" past which the severity of fouling becomes independent of sodium. Inevitably, there are complicating factors: for a given level of sodium, increasing calcium decreases fouling while increasing silicon increases fouling. A fairly complete review of these relationships has been published by E.A. Sondreal and colleagues (21).

A related problem occurs during the fluidized-bed combustion of low-rank coals. The fluidized-bed combustion of high-sodium coals, particularly in silica beds, leads to the formation of agglomerates of ash and bed material. The agglomerates decrease heat transfer and fluidization quality of the bed; in severe cases the formation of

agglomerates requires the combustor to shut down. Research on
agglomerate formation indicates similarities to the ash fouling
mechanism. First, a sulfated layer forms on aluminosilicate ash
particles. Then grains are loosely compacted and held together by a
cement of sulfated aluminosilicates. Eventually the grains are
bonded by a partially melted and recrystallized matrix in which
melilite phases have been identified (22).
    The slagging behavior of coal ash is important in cyclone furn-
aces and in slagging gasifiers. Since the composition of ashes from
low-rank coals differs from that of bituminous coals, empirical
correlations relating viscosity - temperature behavior to composition
developed from bituminous coal slag experience often fail when extended directly to low-rank ash slags. Research on the viscosity
behavior of slags from low-rank coal ash is discussed in Chapter 12
of this book. The data show once again a key role for sodium. It
has also been shown that for two slags having otherwise very comparable compositions the slag with the higher sodium content will
have the lower viscosity at a given temperature (23). Furthermore,
melilites have been identified in the quenched slags from a number of
viscosity tests.
    A common thread runs throughout the behavior of low-rank coal
inorganics in fouling, agglomerating, and slagging phenomena: the
important role of sodium and the formation of melilite phases. As
yet no one has developed a "grand synthesis" of a unified theoretical
explanation of these phenomena. It seems very likely that such a
synthesis is possible, and we may hope that it will be forthcoming in
the not-too-distant future.

Acknowledgments

Special appreciation is due to Edwin Olson and John Diehl for the
data on methoxy content, and to George Montgomery for the ESCA spectrum. Thanks are due to several participants in Associated Western
Universities' research participation programs, particularly David
Kleesattel for separation of the lithotypes of Beulah lignite and for
providing Figure 2, Kathleen Groon for the PDSC thermograms, and Paul
Holm for chemical fractionation and carboxylate functional group
data.

This report was prepared as an account of work sponsored by the
United States Government. Neither the United States nor any agency
thereof, nor any of their employees, makes any warranty, express or
implied, or assumes any legal liability or responsibility for the
accuracy, completeness, or usefulness of any information, apparatus,
product, or process disclosed, or represents that its use would not
infringe privately owned rights. Reference herein to any specific
commercial product, process, or service by trade name, mark, manufac-
turer, or otherwise, does not necessarily constitute or imply its
endorsement, recommendation, or favoring by the United States Government or any agency thereof. The views and opinions of authors
expressed herein do not necessarily state or reflect those of the
United States Government or any agency thereof.

Literature Cited

1. "Low-Rank Coal Study. Volume 1. Executive Summary," U.S. Department of Energy Report DOE/FC/10066-TI (Vol. 1), 1980.
2. "Low-Rank Coal Study. Volume 2. Resource Characterization," U.S. Department of Energy Report DOE/FC/10066-TI (Vol. 2), 1980.
3. Banmeister, T. "Standard Handbook for Mechanical Engineers"; McGraw-Hill Book Co. Inc.: New York, 1967; p. 7-4.
4. Tillman, D.A.; Rossi, A.J.; Kitto, W.D. "Wood Combustion"; Academic Press; New York, 1981; p. 42.
5. Went, F.W. "The Plants"; Time-Life Books: New York, 1963, p. 51.
6. Sondreal, E.A.; Willson, W.G.; Stenberg, V.I. Fuel 1982, 61, 925-38.
7. Benson, S.A.; Schobert, H.H., in "Technology and Use of Lignite," U.S. Department of Energy Report GFETC/IC-82/1, 1982; pp. 442-70.
8. Stein, S.E., in "New Approaches in Coal Chemistry"; Blaustein, B.D.; Bockrath, B.C.; Friedman, S., Eds.; ACS SYMPOSIUM SERIES No. 169, American Chemical Society: Washington, D.C., 1981; p. 104.
9. Farnum, S.A.; Farnum, B.W. Anal. Chem. 1982, 54, 979-85.
10. Schafer, H.N.S. Fuel 1970, 49, 197-213.
11. Schobert, H.H.; Montgomery, G.G.; Mitchell, M.J.; Benson, S.A. Proc. 9th Int. Conf. Coal Sci. 1983, pp. 350-3.
12. Zerda, T.W.; John, A,; Chmura, K. Fuel 1981, 60, 375-8.
13. Miller, R.N.; Given, P.H., in "Ash Deposits and Corrosion Due to Impurities in Combustion Gases"; Bryers, R.W., Ed.; Hemisphere Publishing Corp: Washington, D.C., 1977; pp. 39-50.
14. Schobert, H.H.; Benson, S.A.; Jones, M.L.; Karner, F.R. Proc. 8th Int. Conf. Coal Sci. 1981, pp. 10-15.
15. Suuberg, E.M.; Peters, W.A.; Howard, J.B. Ind. Eng. Chem. Process Res. Devel. 1978, 17, 37-46.
16. "Low Temperature Carbonization Assay of Coal in a Precision Laboratory Apparatus," U.S. Bureau of Mines Bulletin 530, 1953.
17. Petrakis, L.; Grandy, D.W. ACS Div. Fuel Chem. Preprints. 1978, 23(4), 147-54.
18. Retcofsky, H.L.; Hough, M.R.; Clarkson, R.B. ACS Div. Fuel Chem. Preprints. 1979, 24(1) 83-9.
19. Tillman, D.A.; Rossi, A.J.; Kitto, W.D. "Wood Combustion"; Academic Press: New York, 1981, Chap. 4.
20. Honea, F.I.; Montgomery, G.G.; Jones, M.L., in "Technology and Use of Lignite," U.S. Department of Energy Report GFETC/IC-82/1, 1982; pp. 504-45.
21. Sondreal, E.A.; Tufte, P.H.; Beckering, W. Combustion Sci. Tech. 1977, 16, 95-110.
22. Benson, S.A.; Karner, F.R.; Goblirsch, G.M.; Brekke, D.W. ACS Div. Fuel Chem. Preprints 1982, 27(1), 174-81.
23. Schobert, H.H.; Diehl, E.K.; Streeter, R.C., in "Low-Rank Coal Basic Coal Science Workshop," U.S. Department of Energy Report CONF-811268; 1982, pp. 187-96.

RECEIVED March 5, 1984

4

Resources, Properties, and Utilization of Texas Lignite: A Review

T. F. EDGAR[1] and W. R. KAISER[2]

[1]Department of Chemical Engineering, The University of Texas, Austin, TX 78712
[2]Bureau of Economic Geology, The University of Texas, Austin, TX 78712

Texas lignite is now a major energy source in
Texas because of its proximity to principal demand
centers and the Gulf Coast industrial complex and
the growth of demand for electricity in the
Southwest. During the past ten years much research
and development on the assessment of lignite
resources, their chemical and physical properties,
and the utilization of lignite in a wide range of
applications areas has taken place. While direct
combustion is currently the dominant utilization
mode, most of the R&D activity has focused on
newer, alternative technologies such as gasification, fluidized bed combustion, and liquefaction.
In this paper we review recently published
investigations on Texas lignite dealing with the
above subjects, summarizing the key results, and
make some assessments about its future utilization
potential.

Lignite is a major energy source in Texas; production has
increased from 2 to 35 million metric tons per years between
1970 and 1983. Texas now ranks 6th among U.S. coal-producing
states and is the leading lignite-producing state. All of the
state's mines are surface mines which are used mainly for mine-
mouth electric-power generation. Total lignite production is
projected to be 72 million metric tons/year in 1990, corresponding to 13,659 MW$_e$ of installed generating capacity. The large
increase in lignite usage has largely been stimulated by the
rapid escalation in the price of natural gas from $0.15 to $4.50
per million Btu ($.14 to $4.27 per gJ) in the past 15 years.
Natural gas was the principal fuel used for electricial production prior to 1975.

The so-called energy "crisis" which occurred in the mid-
1970's also generated a great deal of interest in uses of lignite

0097-6156/84/0264-0053$07.75/0
© 1984 American Chemical Society

other than for electric power production. A number of laboratory
and pilot scale tests using Texas lignite as a feedstock for
gasification and liquefaction processes have been performed to
develop substitute fuels for oil and gas; in addition, alternative coal combustion technologies such as fluidized bed
combustion have been studied with Texas lignite. These development activities have been the largest impetus to obtaining
fundamental information on Texas lignite; at present only a
modest data base on physical and chemical properties of Texas
lignite exists. This should be contrasted to the significant
long-term effort carried out by the University of North Dakota
Energy Research Center to characterize North Dakota lignite and
evaluate its potential utilization. There has been no systematic effort on the part of the federal or state government to
obtain such information for Texas lignite.

    In this paper we first discuss the geology of Texas lignite
and give some general information about variation of measured
properties. Next we cover the investigations which have focused
on Texas lignite composition: petrology, mineral matter and ash
properties, trace elements, and sulfur compounds. Chemical
reaction properties--kinetics studies on pyrolysis, gasification, oxidation, and liquefaction--are also reviewed. This is
followed by a summary of pilot-scale projects on various
gasification, combustion, and liquefaction processes. Finally
we report some economic studies on utilization of Texas lignite,
which give some indication of how Texas lignite might be employed
commercially in the future.

Resources of Texas Lignite

Stratigraphy. Texas lignite occurs in three Eocene (lower
Tertiary) geologic units - the Wilcox Group, Yegua Formation,
and Jackson Group - deposited in three ancient depositional
systems - fluvial, deltaic, and strandplain/lagoonal. Cyclic
deposition is the fundamental style of sedimentation in the
Texas Eocene. The motif is an alternation of regressive, thick
fluvial-deltaic units such as the Wilcox and transgressive, thin
fossiliferous , marine units such as the Weches (Figure 1).
These units outcrop as belts in the inner Texas Coastal Plain
and extend deep into the subsurface dipping 1° to 2° Gulfward
until, for example, the Wilcox is found at 3,000 m. or more
below the surface at Houston (Figure 2). The Wilcox and Jackson
Groups are the most important lignite-bearing units. Depositional setting has been established from regional, subsurface
lithofacies and lignite-occurrence maps made from more than
4,500 electrical and induction logs. A facies-dependent
exploration model has been developed for Texas lignite (1) that
facilitates resource estimation where data are meager (2).
    In the Wilcox Group of east Texas, lignite at exploitable

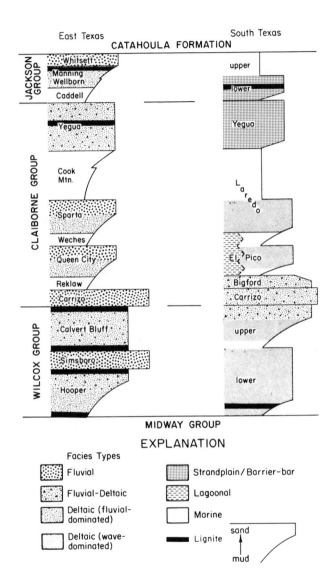

Figure 1.   Cyclic deposition in the Texas Eocene and the occurrence of lignite (modified from (2)).

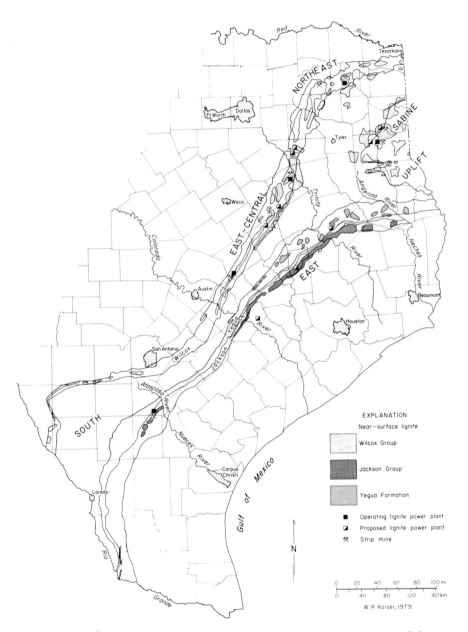

Figure 2. Distribution of near-surface lignite in Texas (1).

depths of 6 to 300 m. is a component facies of ancient fluvial depositional systems whereas lignite in the Jackson Group is of deltaic origin (1-4). In south Texas lignite is associated with strandplain/lagoonal and wave-dominated delta sediments (Figure 1).

Resources. Near-surface lignite is defined as that under less than 60 m. of cover and deep-basin, under more than 60 m. of cover. Near-surface resources are set at 21.2 billion metric tons (1). These resources occur primarily in the Wilcox (71%) and secondarily in the Jackson (22%) and Yegua (7%). Geographically over 90% of the resources are found north of the Colorado River or in east Texas (Figure 2).

Deep-basin resources (seams > 0.6 m. thick and lying between 60 and 1500 m.) were initially estimated by Kaiser (5) at approximately 90 billion tons. However, thick seams at much less than 1500 m. are the economic targets for in situ gasification and deep surface mining. These deep-basin resources (seams > 1.5 m. thick and lying between 60 and 600 m. deep) have been estimated more recently (1) at approximately 31.2 billion metric tons.

Mapping done to date has shown deep-basin resources to occur primarily in the Wilcox and secondarily in the Jackson Groups; deep lignite in the Yegua Formation is relatively unimportant (1). With respect to tonnage, the Wilcox is clearly superior to the Jackson; however, Jackson resources are large enough to support in situ gasification and have more favorable hydrogeological settings. Inspection of lignite occurrence maps shows a geographic distribution similar to that of near-surface lignite (Figure 3). In 1982 and 1983 the Texas Bureau of Economic Geology carried out regional exploratory drilling, logging, and coring of deep-basin Wilcox lignite as well as regional studies of its hydrogeological setting (6).

Quality, seam thickness, continuity. The quality of Texas lignite on an as-received basis ranges from 9.98×10^6 to 17.87×10^6 J/kg (4,300 to 7,700 Btu/lb), 20 to 40% moisture, 10 to 40% ash, and 0.5 to 2.5% sulfur (Table I). The best quality lignite and largest individual deposits occur in the Wilcox Group north of the Colorado River (Fig. 2). On a dry, ash-free basis, the C content indicates a coal rank borderline between lignite and subbituminous. Deep-basin samples from the Sabine uplift area are slightly richer in C (74.3%) than those of east-central Texas (71.3%). Seams are typically 0.6 to 3 m. thick (range from 0.6 to 6.6 m. thick), tabular to lenticular, and continuous for 3.2 to 24.1 km. (2 to 15 miles). Generally, Wilcox lignite north of the Trinity River is of poorer grade and in lenticular and thinner seams (commonly 0.6 to 1.2 m.) than that between the Trinity and Colorado Rivers and on the southern

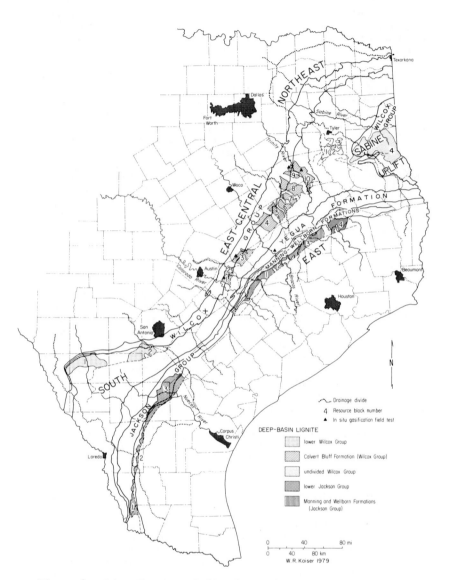

Figure 3. Distribution of deep-basin lignite in Texas (1).

Sabine Uplift (semi-circular outcrop area next to Louisiana),
where tabular and thicker seams (0.9 to 3.6 m. thick) are
present. The poorest grade lignite occurs in the Jackson Group.
Seams are numerous and range from 0.6 to 3.6 m. thick with many
less than 1.5 m. thick. Thicker seams are commonly 1.2 to 2.4
m. thick, tabular, and laterally continuous for up to 48.3 km.
(30 miles).
      Differences in quality and seam continuity are correlated
with ancient depositional environment. Lenticular, moderately
high-ash, low-sulfur lignites of the Wilcox north of the Trinity
River reflect accumulation high on an ancient alluvial plain.
Wilcox stream courses were closely spaced and enclosed relatively small interchannel basins. Thus, associated fresh-water
peat swamps were areally restricted and vulnerable to flood
events and accompanying sediment influx. On the other hand,
south of the Trinity River and on the Sabine Uplift lignites
accumulated lower on the alluvial plain where interchannel
basins were larger, temporally more stable, and less vulnerable
to flood events. In contrast, laterally continuous and moderately high-sulfur lignites of the Jackson Group reflect accumulation in marshes or swamps on the lower delta plain or strand-
plain. Upon abandonment, areally extensive blanket peats formed
and, because of their more seaward sites of accumulation, have
higher sulfur content.

Lignite Properties

Petrography. Few studies on the petrography of Texas lignite
are available (7-10). Recently Jones (10) reviewed these studies
plus others of limited scope and analyzed the petrography of
thirteen deep-basin channel samples taken from cores in the
Wilcox Group of east and east-central Texas. Humodetrinite is
the dominant maceral subgroup and many of the huminites have
undergone partial or complete gelification. The liptinite
content is high and can exceed 30% in some samples; much of it
occurs as a fine-grained matrix. Seams with less liptinite tend
to contain more inertinite. The large amount of attrital
material suggests that Wilcox peat swamps had relatively well-
oxygenated water, allowing extensive bacterial decay. In the
drier swamps fires contributed to the fusinization of partially
coalified peat, whereas in the wetter swamps sapropelic conditions were periodically established as indicated by the presence
of alginite.

Mineral matter. Guven and Lee (11) have performed a characterization of the mineral matter found in several deep-basin lignite
seams. Using core samples from Shelby, Nacogdoches, Panola, and
Rusk Counties in east Texas (Wilcox Group), mineral matter has
been processed using low temperature ashing (LTA) procedures,

TABLE I. Proximate analyses (as-received and dry basis).

	Wilcox east-central	Wilcox northeast	Wilcox Sabine Uplift	Wilcox mean	Jackson east	Jackson south	Jackson mean	Yegua east
			As-received basis, %					
Moisture	32	33	33	33	41	23	37	37
N	83	79	113	275	169	55	224	29
Volatile matter	29	27	27	28	23	22	23	26
N	83	79	113	275	169	55	224	29
Fixed carbon	24	25	25	24	15	13	14	19
N	83	79	113	275	169	55	224	29
Ash	15	15	15	15	21	42	26	18
N	83	79	113	275	169	55	224	29
Sulfur	1.0	0.8	0.9	0.9	1.3	1.9	1.4	1.0
N	76	78	113	267	174	55	229	29
Btu/lb	6,593	6,499	6,441	6,504	4,729	3,972	4,547	5,761
N	83	79	113	275	174	55	229	29

Dry basis, %

Volatile matter	43	41	41	41	40	28	37	41
N	65	78	116	259	153	48	102	29
Fixed carbon	35	37	37	37	25	17	24	31
N	65	78	116	259	153	48	201	29
Ash	22	22	22	22	34	54	39	28
N	65	78	116	259	153	48	201	29
Sulfur	1.4	1.2	1.4	1.3	2.3	2.2	2.2	1.6
N	65	78	113	256	261	48	309	29
Btu/lb	9,715	9,760	9,764	9,750	8,222	5,077	7,733	9,166
N	65	78	116	259	261	48	309	29

[a]Based on percentage of resources.

N = number of analyses.

Note: Analyses do not represent a random sample.

modified from ref. (1)

followed by analysis with x-ray diffraction and analytical electron microscopy.

Quartz, bassanite, kaolinite, and halloysite are the major minerals in these lignites. Bassanite was probably formed from gypsum by dehydration during the ashing process. Pyrite, illite, smectite, chlorite, barite, and feldspars generally occur in smaller quantities. The silica phase consists mostly of authigenic (formed during coalification) microquartz and some detrital quartz. Similarly kaolinite occurred mostly as authigenic vermicular books in the LTA; some additional kaolinite is of detrital origin. Halloysite, gypsum, pyrite, and barite were all believed to be authigenic. The above assemblage of minerals (microquartz + halloysite + kaolinite + pyrite) is probably due to alteration of detrital silicates (such as feldspars and clay minerals) in the acidic peat and coal environment.

The mineralogical compositions of the LTA's from lignites in Shelby, Nacogdoches, and Rusk Counties appear to be rather similar. The mineral matter in the lignites of Panola County is different in that large amounts of detrital clays such as illite and smectite, and rather low quantities of bassanite and pyrite are present.

Trace elements and major oxides. It is well known that most lignites contain highly reactive functional groups--primarily carboxylic acids and hydroxyls--which can form ion-exchangeable bonds with metal cations. These ion exchange sites are found both within the organic structure of the lignite "molecule" and in illitic and smectitic clays. Because of the greater number of available bonding sites in lignites, ion-exchange reactions and chelate formation are much more common than in higher rank coals. Cations forming chelates are more strongly held and are less likely to be lost. These phenomena generally cause some trace element species to be enriched in lignite relative to other coals.

Gulf Coast lignites, in general, based on a limited data set of 38 lignite samples, were found to have higher levels of B, La, Pb, Se, U, Y, and Zr than coals from other U.S. coal basins (12). White et al. have recently compiled all known trace element data for Texas lignite in a TENRAC report (13). They statistically summarized data on 163 samples from 14 Texas counties. Among 32 trace elements and 10 major elements only B, Se, Y, and Zr contents were higher than typically found in other U.S. coals. Wilcox lignite was found to contain more Ge and Se and less Be, Cs, Ga, Li, Mn, Nb, Rb, and Y than Jackson lignite. For all other elements no significant differences between the two lignites were found.

This study also presented some limited data on partitioning of trace elements during combustion. The partitioning of individual trace elements between bottom ash, fly ash, and uncollected

gaseous and fine particulate emissions is significantly different
than occurs for the major elements in coal ash (Al, Si, Ca, Fe,
K, Mg, Na,and Ti). It was reported that elements enriched in
the fly ash/flue gas emissions were Ag, As, B, Cd, Cl, Cr, Cu,
F, Hg, Mo, Ni, Pb, S, Sb, Se, and Zn. The comparison of partitioning data previously developed for bituminous and subbituminous coals with samples of lignite, bottom ash, and fly ash
from two Texas lignite-fired power plants showed great variability in results. Therefore few conclusions regarding trace
element partitioning as a function of coal rank can be drawn at
this time.

Tewalt and others have extensively analyzed 85 east Texas,
deep-basin lignite samples collected from 18 different seams and
five stratigraphic horizons in the Wilcox Group (14). In the
Sabine Uplift area, among the trace elements of greatest
environmental and health concern in coal (As, Pb, Se, B, Hg, Cd,
and Mo), only the mean of Se (10.2 ppm) exceeds that of the
average for U.S. lignite (5.3 ppm), whereas As is less (3.4 vs.
6 ppm). Uranium averaged less than 1 ppm (0.8) compared with
2.5 ppm nationally. In east-central Texas, the means for Se,
As, and U are 6.8, 9.0, and 1.7 ppm, respectively. Thus, Se is
slightly lower and As and U slightly higher than in the Sabine
Uplift samples. In both areas, the means for Pb, B, Hg, Cd, and
Mo are similar to the averages for U.S. lignite.

Analyses of high-temperature ash (Table II) for different
Texas lignites show near-surface Wilcox lignite to have the
lowest Na_2O content and the highest CaO, MgO, and Fe_2O_3 contents.
Jackson lignites are high in SiO_2 and Al_2O_3, reflecting their
higher ash (clay mineral) content. Deep-basin Wilcox ashes are
markedly lower in SiO_2 and higher in Na_2O, SO_3, and CaO than are
ashes from near-surface lignites. The most striking difference
is in Na_2O content, which increases with depth and is correlated
with Na^{+2} content in the ground water (6). Shelby and Panola
County lignites are particularly rich in Na_2O, commonly exceeding 10% in deeper seams.

Zingaro (15) evaluated five lignite cores from Zavala,
Karnes, Atascosa, and McMullen counties in south Texas and from
Freestone County in east-central Texas. Nineteen elements were
included in the analysis (Ag, As, B, Ba, Be, Cd, Co, Cr, Cu, Hg,
Mn, Mo, Ni, Pb, Se, Ti, V, Zn, and F). B, As, Se, and, to a
lesser degree, Pb were the only elements found in concentrations
greater than in typical crustal rocks. Huang and Chatham (16)
and Mohan et al. (17) characterized the uranium content in
lignite samples from the uranium-rich upper Jackson Group.
Huang and Chatham found uranium contents much higher (up to 7800
ppm) at the interface with vertically adjacent sandstone and
shale strata than in the central part of the lignite stream.
Clark, et al. (18) evaluated the arsenic and selenium content of
two lignite cores taken from Freestone County and found arsenic

TABLE II. Ash composition and fusion temperatures, near-surface lignite ([1])

	Wilcox east-central	Wilcox northeast	Wilcox Sabine Uplift	Jackson east	Jackson south[a]	Yegua east
SiO_2	48.53	44.53	40.51	57.58	62.88	53.64
Al_2O_3	17.12	11.39	16.28	20.26	17.49	14.96
TiO_2	1.09	1.14	0.89	0.81	0.82	0.79
CaO	13.20	13.74	12.51	7.49	4.82	9.89
MgO	2.30	2.55	3.03	1.13	0.72	1.81
Fe_2O_3	6.23	7.39	11.46	3.20	2.83	6.61
Na_2O	0.38	0.56	0.80	0.86	3.10	1.23
K_2O	0.84	0.63	0.70	0.84	1.96	1.24
SO_3	9.35	10.87	11.81	7.10	4.62	8.87
P_2O_5	0.11	0.06	0.10	0.05	0.06	0.06
Number of analyses (oxides)	26	21	35	116	design	16

Base/acid[b]	0.34	0.44	0.49	0.17	0.17	0.30
SiO_2/Al_2O_3	2.8	3.9	2.5	2.8	3.6	3.6
Initial deformation	2,226	2,186	2,108	2,240	2,120	2,149
Softening (H=W)	2,261	2,280	2,168	2,351	2,386	2,200
Hemispherical (H=½W)	2,315	2,284	2,215	2,378	2,430	2,256
Fluid	2,357	2,388	2,393	2,466	>2,700	2,475
R	131	202	285	226	> 580	326
Number of analyses (temperature)	27	27	26	32	design	14

Oxides in weight percent.
Fusion temperatures in reducing atmosphere, °F.
a San Miguel lignite.

b $\dfrac{\text{base}}{\text{acid}} = \dfrac{CaO+MgO+Fe_2O_3+Na_2O+K_2O}{SiO_2+Al_2O_3+TiO_2}$

R = range between initial deformation and fluid temperatures.

to occur fairly uniformly throughout the vertical span of the
core while selenium values varied markedly. Washability tests
and efforts at chemical fractionation suggested that both
elements are widely dispersed throughout the lignite, occurring
mainly as fine-grained minerals and in organic assocation.

Washability characteristics. Because Texas lignite has a
moderate sulfur content, there is some interest in its potential
for desulfurization using conventional cleaning techniques. The
sulfur content of Texas lignite varies from about 1.0 to 2.5% on
a dry basis (see Table III). The distribution of sulfur between
pyritic and organic forms is highly variable; generally the
sulfur is organic in nature (usually 70% or higher), although
some mines (e.g., Sandow near Rockdale, Texas) show 50% pyritic
sulfur. Cavallaro and Baker (19) performed an experimental
study on washability characteristics of Arkansas and Texas
lignites, with 7 samples from Texas. Physical coal cleaning
provided significant ash reduction, whereas the reduction in
sulfur content was not marked. Cavallaro and Baker found that
the sodium in Texas lignite was ion-exchangeable with a calcium
solution (37-91% sodium oxide reduction in the ash).

Chemical Reaction Properties

Pyrolysis. All of the Texas lignite pyrolysis data reported in
the literature have been based on slow pyrolysis rates (e.g.,
3-10°C/min). Goodman et al. (20) performed an early study (1958)
on the effect of final carbonization temperature on the yields
from several lignites, one of which was a Wilcox lignite. More
recently Edgar et al. (21) reported a series of atmospheric
pressure studies to evaluate the effects of other carbonization
parameters (heating rate, particle size) on carbonization yield
and decomposition rate.

Oxidation. Tseng and Edgar (22) have studied the combustion
behavior of four Texas lignite chars. Using a thermogravimetric
analyzer, they found that below 550°C, the reaction order of all
lignite samples was 0.7 (with respect to oxygen partial pressure)
and the activation energies ranged from 27.4-27.9 kcal/gmole.
For higher temperatures, the same kinetic relationships hold
although the influence of pore diffusion and external mass
transfer become quite pronounced and are the rate-controlling
steps. Tseng and Edgar found that the pore-diffusion controlled
regime of reaction was quite narrow for lignite char. They also
investigated the effect of pyrolysis temperature on the char
reactivity, finding that char pyrolyzed at 800°C was 3.5 times
more reactive than char pyrolyzed at 1000°C. Tseng and Edgar
developed general correlating equations for coal combustion
which take into account the change in the coal pore structure as
the reaction proceeds.

TABLE III. Forms of sulfur (%).

		Wilcox east-central	Wilcox northeast	Wilcox Sabine Uplift	Wilcox mean	Jackson east	Yegua east
Total sulfur	As received	0.96	0.79	0.94	0.90	1.29	0.99
	N	76	78	113	267	174	29
	Dry	1.41	1.18	1.42	1.35	2.25	1.57
	N	65	78	113	256	261	29
Pyritic sulfur	As received	0.22	0.29	0.26	0.26	0.40	0.50
	N	25	27	30	82	38	14
	Dry	0.34	0.44	0.39	0.40	0.54	0.81
	N	14	29	30	73	141	14
Sulfate sulfur	As received	0.02	0.01	0.03	0.02	0.02	0.02
	N	25	27	30	82	38	14
	Dry	0.01	0.01	0.05	0.03	0.03	0.03
	N	14	29	30	73	141	14
Organic sulfur[a]	As received	0.72	0.49	0.65	0.62	0.87	0.47
	N	25	27	30	82	38	14
	Dry	1.06	0.73	0.98	0.92	1.68	0.73
	N	14	29	30	73	141	14

[a] determined by difference; N = number of samples
modified from ref. (1)

Gasification. The only study on gasification kinetics of Texas
lignite has been performed by Bass (23). Using a differential
reactor, he obtained rate data at 700°C and for pressures ranging
from 61.6 to 225.9 kPa. Rate equations as a function of steam
partial pressure and carbon conversion were developed.

Liquefaction. In this volume Philip et al. (24) have discussed
laboratory studies on the liquefaction behavior of Texas lignite,
hence we do not cover this subject here.

Desulfurization. Chemical desulfurization studies using
H_2O_2/H_2SO_4 and H_3PO_4 have been carried out by Vasilakos et al.
(25). Comparisons with eastern bituminous coals were reported
in this work.
     A related topic of interest for Texas lignite is sulfur
retention by ash during pyrolysis and combustion. Sulfur
retention influences the design of emission control facilities
for combustion and gasification plants utilizing lignite. In
fact a demonstration scale gasifier using Texas lignite has been
permitted to operate without H_2S scrubbers because of laboratory
results indicating high sulfur removals by ash during pyrolysis
and gasification (26). Christman et al. (27) carried out a
series of laboratory-scale experiments (10 g samples) on three
Texas lignites, where the fate of sulfur was analyzed after
pyrolysis and after combustion. The effect of temperature was
studied, and correlations of sulfur retention with ash content
were developed. They found that both calcium and iron influenced
sulfur retention during pyrolysis (reducing environment), but
only calcium reacted with sulfur in an oxidizing environment,
forming $CaSO_4$. There is an optimum temperature for sulfur
retention (around 800°C) under laboratory combustion conditions;
this correlates reasonably well with results for pilot-scale
fluidized bed combustion using limestone as the desulfurizing
agent. This optimum occurs because of the competing effects of
reaction rate (which increases with higher temperature) and
sintering of the ash, making calcium less accessible to reacting
SO_2.
     In a subsequent study by Kochhar (28) data on the sulfation
rate of lignite ash from gravimetric analyses were obtained
along with physical and chemical analyses of the sulfated ash
samples. The gravimetric analyses revealed some mechanistic
information about the sulfation reaction. Temperature and SO_2
concentration were identified as important operating variables.
Increasing sulfation temperature increased initial rate but also
accelerated pore diffusion/pore plugging resistances (product
layer buildup). The effect of increasing SO_2 concentration was
the same as increasing temperature but much less severe. Powder

x-ray diffraction showed the presence of $CaSO_4$ in every sample
tested. In contrast, no Na_2SO_4 was detected, even though Na was
known to be present in the ash. Scanning electron microscope
photographs of lignite and char revealed that the carbon/organic
matrix phase was well-defined from the ash/mineral matter phase.
Dispersed ash agglomerated into a particle after the carbon
burned out during combustion. X-ray energy dispersive spectroscopy showed calcium in ash to be a highly active sorbent of SO_2.
Kochhar indicated that iron, silicon, and aluminum do not react
with SO_2, either catalytically or by occupying reactive calcium
sites. Aluminum and silicon were detected as forming aluminosilicates, especially at high temperatures.

Utilization

Direct combustion has been used with Texas lignite on a utility
scale since the 1950's. However, in the past ten years there
has been considerable testing of Texas lignite in alternative
energy processes such as fluidized bed combustion, gasification,
and direct liquefaction. We summarize activities in each area
below.

Direct (pulverized coal) combustion. The utilization of Texas
lignite by direct combustion has been successful technically and
economically. Fouling and slagging of boilers burning Wilcox
lignite has not been a serious operating problem (29). The Na_2O
content of Wilcox lignite burned at Big Brown, Monticello, and
Martin Lake is less than 0.5%, whereas the base-acid (B/A) ratio
is less than 0.7, and the SiO_2:Al_2O_3 ratio is less than 3.5
(Table II). Ash contents of Wilcox lignites are moderate, being
generally less than 15%. All these factors combine to account
for minimal ash accumulation on convection surfaces and in gas
passages. Contrary to experience with Wilcox lignite, utilization of south Texas Jackson lignite could pose a serious boiler
fouling problems. Jackson lignite, which is used at the San
Miguel steam electric station, was the highest fouling coal ever
tested at the University of North Dakota Energy Research Center
(30). Fouling was attributed to high Na_2O and ash content.
While the Na_2O content is higher than that in any other Texas
lignite ash (Table II; Jackson South lignite), it is less than
5% and not considered excessive. Recent operating experience at
San Miguel has shown that proper boiler design (low volume heat
release, low gas exit temperature, extra furnace height, wider
tube spacing, and numerous sootblowers) can minimize boiler
operating problems, even for a lignite that some consider to be
the lowest quality coal ever used in a large power station in
the U.S. (31).
A second problem posed by lignite ash in power plants is
the collection of fly ash, which is influenced by the chemical

composition of the ash. Kaiser (29) has discussed some of the
operating problems with electrostatic precipitators (ESP) at Big
Brown and Monticello plants (burning Wilcox lignite); the
reported low collection efficiencies stem mainly from the high
resistivity of the ash (see Edgar (32)). This high resistivity
is probably caused by the low sulfur content of the flue gas.
Poor collection efficiency led the operating company to retrofit
a baghouse in parallel with the ESP at Monticello in order to
meet air quality regulations. Design data on the particulate
control equipment used with Texas lignites have been reported by
Kaiser (29).

Atmospheric fluidized-bed combustion (AFBC). Two large tests of
Texas lignite have been operated in pilot-scale AFBC units. The
alkaline components in lignite ash generally serve as scavengers
for sulfur during combustion, thus enhancing sulfur removal
efficiencies. Mei et al. (33) ran a series of experiments on
Texas lignite in a 19-inch bed, but meaningful conclusions are
difficult to draw from the scattered data because of the number
of uncontrolled variables and non-uniform operating conditions.
Radian Corporation, in conjunction with the Combustion Power
Company, performed tests on a 1.2 m. x 1.2 m. unit, with 360
kg/hr feed rate (34). Long duration tests (120 hr) were
operated to obtain more or less steady operating conditions.
Tests with Texas lignites from both the Wilcox (low sulfur) and
Jackson (high sulfur) Groups showed over 90% sulfur removal with
limestone addition and generally under 50% without limestone
addition. Combustion efficiencies with 20% excess air were
generally higher than 99 percent. No bed material agglomeration
or deposit accumulation was observed. The tests demonstrated
smooth operation with lignite resulting in low SO_2, NO_x, and
particulate emissions. Solid wastes produced were analyzed and
found to conform to the non-hazardous classification under the
Resource Conservation and Recovery Act (RCRA). Based on these
results, AFBC of Texas lignites appears to be technically
feasible for industrial steam generation applications.

At the University of Texas at Austin, AFBC pilot plant
investigations are primarily concerned with studying carbon
combustion efficiency and sulfur emissions in the fluidized bed
combustion of Texas lignite. A pilot plant AFBC unit with
capacity of up to 15 kg of coal per hour has been constructed
and fully instrumented for real-time data acquisition. The unit
has logged over 1000 hours of operating time; tests completed to
date include parametric studies of the effects of average bed
temperature, coal particle feed size, and bed material particle
size on carbon combustion efficiency and sulfur dioxide
emissions. A factorial test matrix covering average bed temperature, excess gas velocity, and excess air has also been
completed (35). Results from test runs indicated that the

various bed operating parameters affect the efficiency of carbon
combustion through changes in the average carbon residence time
in the bed and bed carbon concentration.

The UT-Austin experimental tests have shown that high (98+
%) carbon combustion efficiency is possible in a fluidized bed
combustor burning Texas lignite and operating on a once-through
basis. Future tests incorporating recycle of elutriated material
will be performed to determine the extent to which this increases
carbon efficiency.

Gasification. There has been considerable interest in Texas in
lignite gasification. Both conventional and in-situ gasification
technologies have been under development; in the latter case the
main focus has been the deep-basin Texas lignite discussed
earlier. The only gasifier being operated for commercial
purposes in Texas at present is located at the Elgin-Butler
brick plant in Butler, Texas [26]. This plant uses Wilcox
lignite recovered in a small clay and lignite mine. The gasifier
being used is an air-blown, staged, stirred-bed type and does
not require briquetting of the lignite feed. The unit is
designed to fire 80 tons/day of 13.9×10^6 J/kg lignite, producing
8.3×10^{11} J of 5.6 mJ/m^3 gas per day. The 1982 gas cost
projected by Jones et al. (26) was \$2.94/gJ. The gasification
system consumes tars and oils by recycle to the gasifier, and
the plant has been permitted by the Texas Air Control Board
without extensive sulfur or particulate emission controls (no
scrubber or baghouse is employed). The sulfur emissions are
reduced by the alkaline ash material, as discussed earlier.
Startup of the unit is taking place in 1984.

Recently, Shell reported using Texas lignite in the Shell
Coal Gasification Process (formerly Shell-Koppers Process); the
pilot plant operated at 6 tons/day (36). The results were very
favorable, with 99% conversion at an oxygen/MAF coal ratio of
0.85 kg/kg. Maximum syngas production and efficiency occurred
at a ratio of 0.9. Thermal efficiency (cold gas) increased from
about 76% of the coal heating value (lower heating value basis)
to 78% when the reactor pressure was increased from 2.1 mPa to
2.8 mPa. The increase was mainly due to a reduction in relative
heat loss. Slightly more synthesis gas was made at the higher
pressure. Gasification was achieved under very moderate conditions. These results indicated that Texas lignite is an
excellent feedstock for this process. There are little published
data on use of Texas lignite in commercial gasifiers such as
Lurgi, slagging Lurgi, Koppers-Totzek or Texaco, although several
tests have been operated (e.g., (37)).

Since 1975 there has been a substantial amount of research
and development on underground coal gasification (UCG) for Texas
lignite, stimulated by the large resources of deep-basin lignite
and proximity to industry in the Gulf Coast region which uses

large amounts of natural gas. Texas lignite is well-suited for
UCG, being a reactive coal present in relatively thick seams
(1.5 to 4.5 m.). The lignite deposits are centrally located
with respect to the major energy demand centers in Texas, as
shown in Figure 3, permitting the transport of medium-Btu gas to
demand points for steam raising and process heat in Gulf Coast
manufacturing facilities. Other uses of this gas might be on-
site for power production or manufacture of petrochemicals (such
as methanol). The medium-Btu gas described above would consist
chiefly of carbon monoxide and hydrogen and would require use of
pure oxygen for in situ gasification. Another proposed option
is the separation of carbon dioxide from the product gas and
transport for use in tertiary oil recovery in east Texas and
Louisiana. By-product CO_2 recovery considerably enhances the
economics of a UCG facility.
     There have been two industrially-sponsored field programs
operated since 1978:
     1. Basic Resources, Inc. (subsidiary of Texas Utilities),
         near Tennessee Colony (Anderson County).
     2. Texas A&M University plus a consortium of ten companies
         near Alcoa, Texas (Milam County).
The results of the two field programs have been reviewed by
Edgar (38). Basic Resources has a new demonstration project
slated for Lee County which is in the planning stages. This
project represents an extension of Basic Resources' previous
tests (at Big Brown and Tennessee Colony) and will utilize
multiple gasification paths with oxygen injection with a thermal
output of approximately 15 MW_e. The field program will be
carried out in Wilcox Group lignite at a depth of about 150 m.
In 1984 the company received approval for the test by the
appropriate regulatory agencies in Texas.
     Major laboratory and modeling research programs have been
in operation at the University of Texas at Austin and at Texas
A&M University since 1974. The UT-Austin project has emphasized
the development of mathematical models for quantifying the
behavior of an underground coal gasification system, such as for
predicting the chemical composition of the product gas, the
fraction of coal that will be recovered with a particular well
pattern, and the environmental impact, such as subsidence and
aquifer contamination. These modeling activities have been
interfaced to a large experimental program to obtain reaction
properties of the lignite as well as physical and mechanical
properties of the associated overburden (38). Recent results
are discussed in a report to the U.S. Department of Energy (39).
     At Texas A&M the field project is now dormant but geotech-
nical investigations are proceeding. Hoskins and Russell at
Texas A&M are carrying out a project to determine how the
in-situ mechanical properties of the coal and overburden
materials at a field site in Rockdale, Texas, influence the
performance of an in situ gasification experiment.

Liquefaction. Testing of the Exxon Donor Solvent process was
carried out prior to 1982, and a comparison among three coal
feedstocks (Illinois bituminous, Wyoming subbituminous, and
Texas lignite) has been reported (40). For the three coals, the
optimum reactor residence time was 25 to 40 minutes. Longer
residence times caused cracking of liquids to gases. While the
lower rank coals gave higher conversions and demonstrated
enhanced reactivity, the liquid yield for the bituminous coal
was slightly superior to the low rank coals, when the yield from
Flexicoking is included. Another problem with low-rank coals
was high calcium content, which caused deposits on the reactor
walls.

Prospects for Near-Term Utilization

With the growth in energy demand in the Southwest, there will be
a continuing increase in lignite consumed for electric power
generation during the rest of this century. Most of the impetus
for new utilization projects is of course based on economics of
competing technologies. By 1990, alternative uses also will
begin to appear, led by gasification and fluidized bed combustion. Lignite liquefaction will probably not be utilized on a
commercial scale in the next 10 years.

Underground gasification is an attractive technology because
it eliminates the mining operation. Economic studies (38)
indicate that medium Btu gas from UCG can be competitive with
synthesis gas from reforming natural gas, if the unregulated
natural gas price rises in the future. However, a number of
technical issues require future research and development involving field testing. Future goals should include relating the
process characteristics to geological conditions (site selection), developing a valid predictive capability for roof collapse
and surface subsidence, as well as methods for its control;
minimizing negative environmental impact of ground water
pollutants and predicting the complex physicochemical and
biological phenomena which occur underground after gasification
has ceased; and predicting the quality and quantity of gas
produced and the recovery of coal for a given injection rate,
injection gas composition, well spacing, water influx, and seam
thickness.

Another gasification alternative under consideration is the
use of entrained-bed gasification with Texas lignite, such as
the Shell or Texaco processes (36). Such a gasifier could be
used to produce synthesis gas or be used in conjunction with a
combined-cycle power generation facility. A recent EPRI report
prepared by Fluor Engineers (41) presented the results of a
detailed engineering and economic evaluation of such a power

generation facility. A Shell-based integrated gasification-combined cycle power plant had a 30-year constant dollar levelized bus-bar cost of electricity of 36.77 mills (1981)/kWh. A steam power plant firing Illinois No. 6 coal had an estimated electrical cost of 43.86 mills (1981)/kWh.

EPRI contracted an economic study on the use of Texas lignite in atmospheric fluidized bed combustion (42), and projected that the costs of AFBC would be less than those for pulverized combustion plus flue gas desulfurization. However, while AFBC appears feasible for small-scale units (less than 50 MW_e), there is some doubt that large scale utility AFBC systems can operate successfully. Part of this problem stems from the inadequacy of "bubbling bed" design normally used in AFBC; newer designs, such as the circulating bed, offer more promise for commercial application and are being tested presently (32).

Literature Cited

1. Kaiser, W. R.; Ayers, W. B., Jr.; LaBrie, L. W. "Lignite Resources in Texas"; University of Texas at Austin, Bureau of Economic Geology Report of Investigations No. 104, 1980.
2. Kaiser, W. R. "Lignite Depositional Models, Texas Eocene: A Regional Approach to Coal Geology"; In Proceedings Basic Coal Science Workshop; Schobert, H. H., ed.; U.S. Department of Energy, Grand Forks Energy Technology Center, and Texas A&M University, 1982.
3. Kaiser, W. R.; Johnston, J. E.; Bach, W. N. "Sand-body Geometry and the Occurrence of Lignite in the Eocene of Texas"; University of Texas at Austin, Bureau of Economic Geology Geological Circular 78-4, 1978.
4. Kaiser, W. R. Proc. 1976 Gulf Coast Lignite Conference; 1978, CONF-7606131, 33-53.
5. Kaiser, W. R. "Texas lignite: Near-surface and Deep-basin Resources"; University of Texas at Austin, Bureau of Economic Geology Report of Investigations No. 79, 1974.
6. Kaiser, W. R. et al. "Evaluating the Geology and Ground-water Hydrology of Deep-basin Lignite in the Wilcox Group of East Texas"; University of Texas at Austin, Bureau of Economic Geology Final Report for Texas Energy and Natural Resources Advisory Council, 1984.
7. Selvig, W. A.; Ode, W. H.; Parks, B. C.; O'Donnel, M. S. "American Lignite: Geological Occurrence Petrographic Composition and Extractable Waxes", Bull. U.S. Bur. Mines, 1950, No. 482.
8. Parks, B. C.; O'Donnell, M. S. "Petrography of American Coals", Bull. U.S. Bur. Mines, 1956, No. 550.
9. Nichols, D. J.; Traverse, A. Geoscience and Man, 1971, 3, 37-48.

10. Jones, C. M. "Petrology of Lignites from the Wilcox Group
    (Lower Tertiary) of East and East-Central Texas"; Bureau of
    Economic Geology, University of Texas, Austin, Texas,
    Open-File Rept., January, 1984.
11. Guven, H.; Lee, L-J. "Characterization of Mineral Matter in
    East Texas Lignites"; Texas Energy and Natural Resources
    Advisory Council, Report EDF-103, Austin, Texas, August,
    1983.
12. White, D. M.; Edwards, L. O.; et al. "Correlation of Coal
    Properties with Environmental Control Technology Needs for
    Sulfur and Trace Elements"; Radian Corporation, EPA
    Contract 68-01-3171.
13. White, D. M.; Edwards, L. O.; DuBose, D. A. "Trace Elements
    in Texas Lignite"; Texas Energy and Natural Resources
    Advisory Council Report EDF-094, Austin, Texas, August,
    1983.
14. Tewalt, S. J. et al. "Chemical Characterization of Deep-
    basin Lignites"; in Kaiser et al. (6).
15. Zingaro, R. A. "Trace Element Characterization in Texas
    Lignites and Liquids Derived Therefrom"; Texas A&M Univer-
    sity, College Station, TX, TENRAC Report No. EDF-060, 1981.
16. Huang, H.; Chatham, J. R. Tenth International Congress on
    Sedimentology, Jerusalem, Israel, 1978.
17. Mohan, M. S.; Zingaro, R. A.; Macfarlane, R. D.; Irgolic,
    K. J. Fuel, 1982, 61, 853-858.
18. Clark, P. J.; Zingaro, R. A.; Irgolic, K. J.; McGinley, A.
    N. Intern. J. Environ. Anal. Chem., 1980, 7, 295-314.
19. Cavallaro, J. A.; Baker, A. F. "Washability Characteristics
    of Arkansas and Texas Lignites"; EPA 600/7-79-149,
    Washington, D.C., June, 1979.
20. Goodman, J. B.; Gomez, M.; Parry, V. F. "Laboratory
    Carbonization Assay of Low-Rank Coals at Low, Medium, and
    High Temperatures"; U.S. Bureau of Mines, Report of
    Investigations 5383, U.S. Department of the Interior, 1958.
21. Edgar, T. F. et al. Proc. of Sixth UCC Symp., 1980,
    IV-9-IV-17.
22. Tseng, H. P.; Edgar, T. F. Fuel, 1984, 63, 385-395.
23. Bass, J. E. M.S.E. Thesis, University of Texas, Austin,
    Texas, December, 1983.
24. Philip, C. B.; Anthony, R. G.; Zhi-Dong, C. In The Chemistry
    of Low-Rank Coals; Schobert, H. H., ed.; American Chemical
    Society, Washington, D.C., 1984.
25. Vasilakos, N. P. "Chemical Coal Beneficiation with Aqueous
    Hydrogen Peroxide/Sulfuric Acid Solutions"; preprints,
    Division of Fuel Chemistry; American Chemical Society; Vol.
    28, Aug. 28-Sept. 2, 1983, Washington, D.C.
26. Jones, E. J.; Hamilton, E. P.; Edgar, T. F. "Application of
    a Sawdust Gasifier to Produce Low-Btu Gas from Lignite at a
    Texas Brick Plant"; Coal Technology Conference, Houston,
    Texas, 1982.

27. Christman, P. G.; Athans, M. P.; Edgar, T. F. AIChE Symp. Series, 1982, 78 (216), 30–40.

28. Kochhar, R. Ph.D. Dissertation, The University of Texas, Austin, Texas, May, 1984.

29. Kaiser, W. R. "Electric Power Generation from Texas Lignite"; Geological Circular 78-3, Bureau of Economic Geology, The University of Texas, Austin, Texas, 1978.

30. Sondreal, E. A.; Gronhovd, G. H.; Kube, W. R. Proc. Gulf Coast Lignite Conference, 1978, CONF-7606131, 100–124.

31. Haller, K. H. "Design for Lignite Firing at the San Miguel Power Plant", 11th Biennial Lignite Symposium, San Antonio, Texas, June 1981.

32. Edgar, T. F. "Coal Processing and Pollution Control", Gulf Publishing Company: Houston, Texas, 1983.

33. Mei, J. S.; Grimm, U.; Halow, J. S. "Fluidized-Bed Combustion Test of Low Quality Fuels - Texas Lignite and Lignite Refuse"; MERC/RI-78/3, 1978.

34. Talty, R. D. et al. "Fluidized Bed Combustion Studies of Texas Lignite"; Technology and the Use of Lignite, GFETC/IC-82/1, 565 (1981).

35. Westby, T.; Kochhar, R.; Edgar, T. F. "Retention of Sulfur by Ash During Fluidized Bed Combustion of Texas Lignite"; AIChE Meeting, Houston, Texas, April, 1983.

36. Heitz, W. L. "Applicability of the Shell Coal Gasification Process (SCGP) to Low Rank Coals"; AIChE Meeting, Atlanta, Georgia, March, 1984.

37. McGurl, G. V.; Farnsworth, J. F. Proc. Gulf Coast Lignite Conference, 1978, CONF-7606131, 141–152.

38. Edgar, T. F. AIChE Symp. Ser., 1983, 79, (226), 66–76.

39. Edgar, T. F.; Humenick, M. J.; Thompson, T. W. "Support Research on Chemical, Mechanical, and Environmental Factors in Underground Coal Gasification"; Final Report to U.S. Department of Energy, Contract DE-AS20-81LC10713, 1984.

40. Mitchell, W. N.; Trachte, K. L.; Zacaepinski, S. "Performance of Low Rank Coals in the Exxon Donor Solvent Process"; Symposium on Technology and Use of Lignite, GFETC/IC-79/1 1979.

41. Hartman, J. J.; Matchak, T. A.; Sipe, H. E.; Wu, M. "Shell-based Gasification-Combined Cycle Power Plant Evaluations"; EPRI Report AP-3129, Palo Alto, California, June, 1983.

42. Burns and Roe. "Conceptual Design of a Gulf Coast Lignite-Fired Atmospherica Fluidized-Bed Power Plant"; EPRI Report FP-1173, Palo Alto, California, September, 1979.

RECEIVED June 21, 1984

PHYSICAL AND CHEMICAL
CHARACTERIZATION

Small-Angle X-Ray Scattering of the Submicroscopic Porosity of Some Low-Rank Coals

HAROLD D. BALE[1], MARVIN L. CARLSON[1], MOHANAN KALLIAT[2,3], CHUL Y. KWAK[2,4], and PAUL W. SCHMIDT[2]

[1]Physics Department, University of North Dakota, Grand Forks, ND 58202
[2]Physics Department, University of Missouri, Columbia, MO 65211

From the small-angle x-ray scattering curves of some American subbituminous coals and lignites, the specific surfaces of the macropores and transitional pores have been calculated. Subbituminous coals contain a larger fraction of transitional pores than are present in lignites. A method is proposed for calculating the distribution of pore dimensions in lignites from the scattering data. This technique provides values of the pore-dimension distribution for average pore dimensions from about 4 to at least 300 Å.

Small-angle x-ray scattering is often useful for investigating submicroscopic structures—that is, structures with dimensions between about 4 and 2000 Å (1). This technique has been employed in several studies of the porosity and other submicroscopic structure of coals (2–9).

We have recently completed a small-angle x-ray scattering investigation of some lignites and subbituminous coals. As we explain in the next section, the specific surfaces of the macropores and transitional pores can be computed from the scattering data. In the concluding section we suggest a technique for obtaining the distribution of pore dimensions in lignites from the scattering curves.

Scattering Measurements, Data Analysis, and Evaluation of Specific Surfaces

Figure 1 is a schematic drawing of a small-angle x-ray scattering system. X-rays from the tube T are formed into a beam by slits and fall on the sample S. A small fraction of the x-rays striking the

[3]Current address: Bell Laboratories, 4500 S. Laburnum Ave., Richmond, VA 23231
[4]Current address: Korea Standard Research Institute, Daedok, Korea

0097-6156/84/0264-0079$06.00/0
© 1984 American Chemical Society

sample are re-emitted, without change of wavelength, in directions different from that of the incoming beam. The intensity of these re-emitted x-rays, which are called the scattered rays, and their dependence on the direction in which they are emitted depend on the structure of the sample. In a scattering experiment, the intensity of the x-rays scattered in different directions is measured, and from an analysis of these data, an attempt is made to obtain information about the structure of the sample producing the scattering. Figure 1 shows a ray scattered at an angle θ with respect to the incoming beam. The scattered radiation is recorded by the detector C.

While there is no universal prescription for analyzing the scattering pattern from an arbitrary sample, we will list some general principles useful for interpretation of scattering measurements. For a sample which has a structure characterized by a dimension a, most information obtainable from scattering measurements will be found at scattering angles θ which satisfy the condition

$$0.1 < ha < 10 \qquad\qquad (1)$$

where $h = (4\pi/\lambda) \sin(\theta/2)$; and λ is the x-ray wavelength. For angles no greater than about 7 degrees, $\sin \theta/2$ can be approximated by $\theta/2$, and so for small scattering angles, h can be considered proportional to θ. According to Inequality (1), for a structure with dimension a, the scattering is determined by the product ha, so that there is an inverse relationship between the size of the structure and the h values at which the scattered intensity from this structure is appreciable. Since the x-ray wavelengths are normally of the order of 1 or 2 Å and thus are of the same magnitude as the interatomic spacing in solids and liquids, Inequality (1) states that the x-ray scattering from structures with dimensions between about 4 and 2000 Å will be observed at scattering angles no greater than a few degrees. Small-angle x-ray scattering thus can be used to study these submicroscopic structures.

X-rays are scattered by electrons, and the small-angle scattering will be appreciable when the sample contains regions in which fluctuations or variations in electron density extend over distances of 4 to 2000 Å. At small angles, the scattering process is unable to resolve structures smaller than about 4 Å, and so in the analysis of the scattering data, the atomic-scale structure can be neglected. For many scattering studies, it is therefore convenient to consider the sample to be composed of two phases, with constant but different electron densities (10).

If this two-phase approximation holds, so that the two phases always are separated by a sharp, discontinuous boundary, and if the minimum characteristic dimension a_m of the structure satisifies the condition $ha_m > 3.5$, the scattered intensity I(h) can be approximated by (7)

$$I(h) = \frac{2\pi\rho^2 (S/M)}{h^4} \frac{M}{A} I_e A \qquad\qquad (2)$$

where ρ is the difference of the electron densities of the two phases, I_e is the intensity scattered by a single electron; S is the total surface area separating the two phases in the sample; M is the mass of the sample; and A is the cross-sectional area of the sample perpendicular to the incident beam. In the outer part of the small-angle scattering curve--that is, when $ha_m \gtrsim 3.5$, the scattered intensity is proportional to h^{-4} and thus to the inverse fourth power of the scattering angle. Moreover, when I(h) has this angular dependence, the magnitude of the scattered intensity is proportional to the specific surface S/M, which is the surface area separating the two phases per unit sample mass.

As we explain in Reference (7), the quantities I_eA and M/A can be evaluated from the x-ray data, and so Equation (2) can be employed to calculate the specific surface S/M from the scattering data for samples with submicroscopic porosity.

We have used small-angle x-ray scattering to determine the specific surfaces of a number of PSOC coals (7,11). As our first x-ray studies of these coals showed (7) that almost all of the small-angle x-ray scattering was due to pores and that the scattering from the mineral matter in the coal was almost or completely negligible, we have interpreted the small-angle x-ray scattering from coals as being due to submicroscopic pores, which are filled with air and for scattering purposes thus can be considered to be empty.

The curves shown in Figure 2 are typical of the scattering curves which we obtained for subbituminous coals and lignites. Just as in the curves shown in Figure 2 of Reference 7, for all subbituminous coals and lignites which we examined, the inner part of the scattering curve, at least for scattering angles between 0.003 and 0.005 radians, was nearly proportional to the inverse fourth power of the scattering angle. In plots in which both axes were logarithmic, the innermost portion of the scattering curves from the subbituminous coals and lignites thus were linear, with slope approximately -4. As the scattering angle increased, the intensity decayed and approached a low but constant value at the largest scattering angles for which we recorded data. In the curves for almost all of the subbituminous coals, there was at least the trace of a shoulder at scattering angles of approximately 0.010 or 0.020 radians. On both sides of the shoulder, the intensity is approximately proportional to the inverse fourth power of θ. (The shoulder often occurs at such small angles that it is not always easy to see the inner region in which the intensity is nearly proportional to the inverse fourth power of θ.) As we explain in Reference 7, we consider the presence of this shoulder to be an indication that in these coals there is a relatively high fraction of transitional pores, with average dimensions of the order of a few hundred Å.

All of our scattering measurements were made on powdered coal samples which had been ground in air and sealed in stainless steel samples cells with Mylar® plastic windows.

We measured the x-ray scattering from Beulah lignite and the PSOC subbituminous coals and lignites on a Kratky camera (12) at the University of Missouri. The scattering data were corrected for background scattering and collimation effects by the procedures outlined in Reference 7. As in our earlier studies of coals, we employed a proportional counter, a nickel-foil filter, and a pulse height analyzer to select x-rays with wavelengths in a narrow band

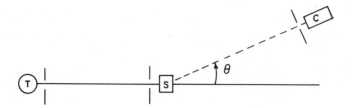

Figure 1. A schematic diagram of a small-angle X-ray scattering
system. The X-rays from the tube T are formed into a beam by the
slits and stride the sample S. Another slit is used to define the
beam scattered at a scattering angle . The scattered X-rays are
detected by the counter C. (Reproduced from Ref. 7. Copyright
1981, American Chemical Society.)

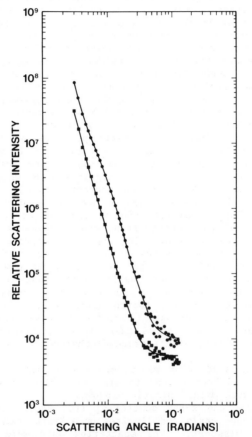

Figure 2. Scattering curves for PSOC 93 lignite (circles) and
PSOC 248 subbituminous coal (squares). The points show the
corrected scattering data, and the curves were obtained from
least-square fits of Equation 3 with n=1.

around 1.54 Å. We also obtained scattering data from Beulah lignite at the University of North Dakota, using a Beeman four-slit collimation system (13) for the outer part of the scattering curve and employing a Bonse-Hart system (14) to record the inner portion. The scattering curves from the two collimation systems were combined to provide composite scattering curves. The Bonse-Hart system enabled us to measure the scattering at angles as small as 0.0003 radians, while with the Beeman and Kratky systems, the smallest accessible angle was 0.003 radians. In addition to the proportional counter and pulse height analyzer employed with the Krarky system to select wavelengths in a narrow interval around the Cu Kα wavelength (1.54 Å), in the Beeman system we also installed a pyrolytic graphite crystal in front of the proportional counter. This crystal appreciably narrowed the band of wavelengths registered by the detector. The interval of wavelengths recorded by the Bonse-Hart system was similar to that obtainable with the Beeman system. Collimation corrections were made by methods analogous to those discussed in Reference (7).

As our earlier investigations of coal suggested that the scattering was due to pores in the coal (7), we analyzed our scattering data under the assumption that the scattering was produced by three classes (15) of pores--macropores, with dimensions of 1 micron or more; transitional pores, which have dimensions of a few hundred Å, and micropores, which were not larger than about 20 or 30 Å.

For the analysis of our scattering data, we made least-squares fits of the equation

$$I(h) = 2\pi\rho^2 I_e A \frac{M}{A} \left[\frac{A_0}{h^{-4}} + \frac{B_{mi}}{(1 + ha_{mi}^2)^2} + \sum_{i=1}^{n} \frac{A_i}{(c_i^2 + h^2)^2} \right] \qquad (3)$$

Each of the terms in the sum in Equation (3), which we have found to be a physically reasonable and convenient expression to describe the small-angle scattering from coal samples, represents the scattering from a group of pores. The form of these terms (3) was suggested by the expression developed by Debye, Anderson and Brumberger for analysis of small-angle scattering data from porous materals (16). The constant A_0 is the specific surface area of the macropores in the sample, and the A_i in Equation (3) are the specific surfaces of the pores in class i. The reciprocals of the c_i are proportional to the average dimensions of this group of pores. The term in Equation (3) proportional to B_{mi} describes the scattering from the micropores. The constant a_{mi} is proportional to the average dimension of the micropores, which were usually so small that in the least squares fits we could consider a_{mi} to be equal to 0.

Although we were able to obtain satisfactory fits of (3) with n = 1 for all of the low-rank PSOC coals discussed below, the scattering data for Beulah lignite obtained with the Bonse-Hart and Beeman systems extended over such a wide interval of scattering angles that we found that n had to be increased to 2 in order to obtain a good fit for this curve.

Corrected scattering curves for PSOC coals 93 and 248 are shown in Figure 2, and Figure 3 is a plot of the corrected data for PSOC coal 86. The scattering curves for the other lignites and sub-bituminous coals were quite similar to the curves for PSOC 93 and PSOC 248, respectively. Figure 4 shows the scattered intensity for Beulah lignite obtained at the University of North Dakota. The curves recorded for Beulah lignite in the two laboratories are in good agreement except at the largest scattering angles, where a higher scattering intensity was recorded at the University of Missouri. Since the measurements in the two laboratories were per-fomred with different samples, we remeasured the scattering from Beulah lignite on both the Beeman and Kratky systems at the Univer-sity of Missouri, using the same sample that was studied at the University of North Dakota. As before, all scattering curves were nearly identical except at the largest scattering angles, where the Kratky system showed a higher intensity than the Beeman systems, although the differences between the curves were smaller than we observed previously. At least part of the remaining difference can be attributed to flourescence scattering from the iron in the minerals in Beulah lignite. Although the graphite crystals on the Beeman systems exclude iron K-fluorescence wavelengths (1.94 Å), the pulse height analyzer employed with the Kratky system can be expected to pass at least a part of this radiation, which, though weak, is essentially independent of the scattering angle and thus is most noticeable in the outer part of the small-angle scattering curves, where the scattering from the pores in the coal is quite weak. We would like to point out, however, that the curves obtained in the two laboratories differed only in the region of the curve which provides information about the micropores.

Values of the specific surfaces of the macropores and transitional pores calculated from least-squares fits of Equation (3) for the PSOC coals and Beulah lignite are given in Table 1. The micropores were so small that we were unable to determine a_{mi} from the fits and instead could calculate only B_{mi}. The c_i^{-1} computed in our fits vary from about 25 Å to 45 Å. Since the c_i, which are proportional to the reciprocals of the average dimensions of the transitional pores, did not appear to vary systematically with the fixed carbon content or any other properties of the coals, we have not included values of the c_i in Table 1. Neither have we listed the b_{mi}, since there is no simple relation between these quantities and the properties of the micropores.

We calculated the specific surfaces shown in Table 1 by an improvement of the procedure described in Reference 7. In our more recent studies of coals, rather than using the mass absorption coef-ficient of carbon, we have computed the mass absorption coefficient of each coal from the elemental composition given by the ultimate analysis. These mass absorption coefficients, which depend quite strongly on the composition and concentration of minerals in the coals, varied from about 7 to 12 cm^2/gm. We also have taken the values of the coal densities from Fig. 2 of Reference (17). This plot shows the coal density as a function of fixed carbon content and thus provides more reliable densities than the approximation we used in Reference (7). The quantity $I_e A$ was calculated from the scattering data for colloidal silica samples by the procedure outlined in Reference 7. The proximate and ultimate analyses of

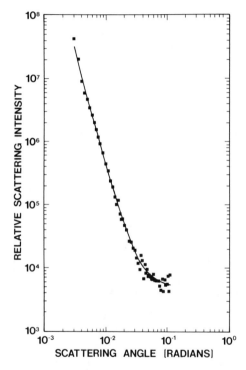

Figure 3.   The corrected scattering curve for PSOC 86 lignite (squares) and the fit of Equation 6 (line).

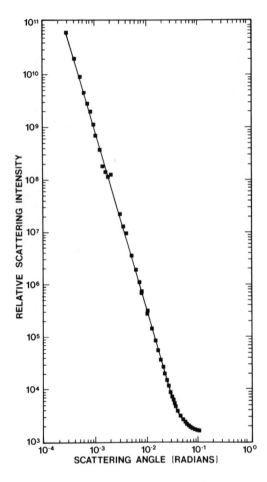

Figure 4. The scattered intensity for Beulah lignite as
measured with the Bonse–Hart and Beeman collimation systems
(squares) at the University of North Dakota. The curve was
drawn from a least-squares fit of Equation 6, and the points
are the corrected scattered intensities.

Beulah lignite are given in Table 2. (Table 2 does not give the ultimate analysis for the mineral matter and thus does not provide

TABLE 1
X-Ray Specific Surfaces for Some Low-Rank Coals

COAL	RANK	STATE OF ORIGIN AND SEAM NAME	X-RAY SPECIFIC SURFACE (m^2/gm)		FIXED CARBON
			Macro-pores	Transi-ional Pores	(Dry, ash-free, Per Cent)
PSOC 64A	Sub-bituminous B	Wyoming	1.2	6.1	72.06
PSOC 86	Lignite	North Dakota Zap	3.1	6.3	
PSOC 92	Lignite	Montana Lower Lignite	2.8	4.6	69.98
PSOC 93	Lignite	Montana Lower Lignite	2.8	2.7	71.95
PSOC 97	Sub-bituminous A	Wyoming Bed #80	3.2	17.8	75.03
PSOC 138	Sub-bituminous C	Texas Darco Lignite	4.3	6.8	74.32
PSOC 240	Sub-bituminous B	Wyoming	2.4	8.7	
PSOC 240A2	Sub-bituminous B	Wyoming Big Dirty	3.2	9.5	
PSOC 242	Sub-bituminous B	Wyoming Dietz	3.3	28	73.87
PSOC 246	Lignite	North Dakota Hagel	2.3	5.8	70.97
PSOC 248	Sub-bituminous A	Wyoming Adaville #1	5.2	39	75.16
PSOC 527	Sub-bituminous A	Wyoming Dietz #2	1.91	12.2	74.92
	Lignite (a)	North Dakota Beulah	2.4	3.4	70.90
	Lignite (b)	North Dakota Beulah	2.0	4.4	70.90

(a) From scattering data measured on the Kratky camera.
(b) From the scattering curves recored on the Bonse-Hart and Beeman systems.

enough data to calculate the mass absorption coefficient, which was obtained from results not included here.) As we had no analyses for PSOC coals 86, 240, and 240A2, we used the mass absorption coefficient calculated for PSOC 246, another North Dakota lignite, in the analysis of PSOC coal 86, and we assumed that the mass absorption coefficients of PSOC coals 240 and 240A2 were the same as for

PSOC 242, which also is a subbituminous coal. Since we did not know
the fixed carbon content of PSOC coals 86, 240, and 240A2, we took
their mass density to be 1.35 gm/cm^3. Uncertainties in the coal
densities probably will not appreciably increase the overall uncertainty in the calculated x-ray specific surfaces, because the
density of low-rank coals changes by only a few per cent for fixed
carbon contents between about 69 and 73% [Ref. 17, Figure 2].

     Because of the many quantities needed to compute the x-ray
specific surface, we cannot claim accuracies greater than about 25
or 50% for the x-ray specific surfaces given in Table 1.

     The total x-ray specific surfaces of Beulah lignite determined
at the University of North Dakota and the University of Missouri
agree within the estimated uncertainty. For the specific surfaces
of the macropores and the transitional pores, the agreement, though
still within the estimate uncertainty, is not so close. This
greater difference may be a result of the fact that the scattering
curves from the Universities of North Dakota and Missouri were
determined by fits of Equation (3) with n = 2 and n = 1,
respectively. The different n values lead to different divisions
between the specific surfaces ascribed to macropores and transitional pores. The difference between the specific surfaces
associated with the macropores and transitional pores calculated
from the scattering curves measured in the two laboratories thus can
be expected to be somewhat larger than the difference between the
total x-ray specific surfaces.

     The transitional-pore specific surfaces listed in Table 1 for
all of the subbituminous coals except PSOC 138 are about 3 to 8
times as large as the corresponding macropore specific surfaces.
The ratio of the area associated with the transitional pores to that
for the macropores is large enough in these subbituminous coals that
at least a trace of an inflection can be seen on careful examination
of the scattering curves. On the other hand, for the subbituminous
Coal 138 and for the lignites, the specific surfaces of the
macropores and the transitional pores are more nearly equal, and no
inflection is evident in the scattering curves.

     Though the fixed-carbon content of PSOC 138 is high enough that
this coal has been classified as subbituminous, it is called Darco
lignite. Since this name suggests that PSOC coal 138 may be rather
similar to a lignite, we are not surprised that the scattering curve
for this coal is so much like the curves we obtained from lignites.

     At the time we wrote the manuscript for Reference (7), we were
unable to relate the high fraction of transitional pores found in
many of the coals to any other property of the coals. We have now
concluded (11) that there are inflections in the scattering curves
only for coals with fixed carbon contents in the interval from about
72% through 83% per cent (dry, mineral-matter free) and that for
these coals, the specific surface of the transitional pores is
appreciably larger than the specific surface of the macropores. As
would be expected from our interpretation of the inflection (11),
the specific surfaces of the transitional pores in all of the
subbituminous coals in Table 1 except PSOC 138 are much larger than
the specific surfaces of the macropores, and there is a tendency for
the total x-ray specific surfaces [i. e., the sum of the specific
surfaces of the macropores and transitional pores, as calculated by

a fit of Equation (3) to the scattering data] to increase as the percentage of fixed carbon (dry, mineral—matter—free) rises. (As we have mentioned previously, PSOC 138 behaves somewhat anomalously.) For the lignites listed in Table 1, the specific surfaces of the transtional pores and the macropores are of the same magnitude, and there are no inflections in the scattering curves from lignites, as we would predict from the concepts introduced in Reference (11), since for lignites the fixed—carbon content is below 72%.

The ability to distinguish between the specific surfaces associated with the macropores and the transitional pores is, we feel, a very useful property of small—angle x—ray scattering especially since few other techniques can identify the contributions of these two pore classes to the specific surface.

Table 2
Proximate and Ultimate Analyses of Beulah Lignite

Proximate Analysis (weight per cent)		
	As Received	Dry, ash—free
Moisture	10.43	
Volatile Matter	37.85	46.71
Fixed Carbon	43.17	53.29
Ash	8.55	
Ultimate Analysis (weight per cent)		
	As Received	Dry, ash—free
Hydrogen	5.04	4.79
Carbon	57.44	70.90
Nitrogen	.69	.85
Sulfur	.04	1.29
Oxygen (1)	27.25	22.19

(1) By difference

The values of the specific surface obtained in studies of coals and many other porous materials depend on the techniques used for the measurement. For example, Gan, Nandi, and Walker (17), in their adsorption studies of a number of PSOC coals, obtained much larger specific surfaces by carbon dioxide adsoprtion at room temperature than by low—temperature adsorption of nitrogen. As we mentioned in Reference (7), the difference may be the result of the fact that carbon dioxide at room temperature can penetrate smaller pores than those which nitrogen molecules can enter at low temperature (18). A comparison of the magnitudes of the x—ray specific surfaces in Table 1 and in Table 1 of Reference 7 with the specific surface values in Figure 1 of Reference (17) shows that the x—ray specific surfaces are much nearer those measured by low temperature nitrogen adsorption than by room temperature carbon dioxide adsorption. As the micropores are too small to satisfy the conditions necessary to give a scattered intensity proportional to the inverse fourth power of the scattering angle, the "x—ray specific surface" includes only contributions from the macropores and transitional pores and does not take account of the specific surface of the micropores.

Estimation of Pore-Dimension Distributions

In the preceding section, we calculated the specific surfaces of the macropores and the transitional pores from the scattering curves. We now would like to suggest a method which we feel may be useful for obtaining at least a rough estimate of the <u>distribution</u> of pore dimensions in low-rank coals.

One of us has recently shown (19) that when the scatterers in a sample are three-dimensional (that is, are <u>not</u> platelets or rods), and if these scatterers have the same shape, scatter independently of each other, and have dimensions described by the dimension distribution function

$$\rho(a) = Ka^{-\gamma} \qquad (4)$$

where K and γ are constants, the scattered intensity $I(h)$ is given by

$$I(h) = I_{oo}h^{-(7-\gamma)} \qquad (5)$$

where I_{oo} is a constant. [The distribution function $\rho(a)$ is defined to have the property that $\rho(a)$ da is proportional to the probability that the observed scattering is produced by a scatterer with a characteristic dimension with a value between a and a + da.] As is explained in Reference (19), for systems of scatterers with the dimension distribution (4), the scattering is given by Equation (5) for all three-dimensional scatterers, <u>regardless of their shape.</u>

In the scattering curves from the lignite samples which we have studied, there is a rather long region in which the scattering is proportional to a negative power of h, as can be seen from the long straight-line portion in the logarithmic plots in Figs. 2, 3, and 4. Although this power-law angular dependence tempted us to use Equations (4) and (5) for analysis of the scattering data from the lignites, we were at first reluctant to employ this technique, because we felt that if the analysis was to be more than a mathematical game, we should have some independent evidence that the distribution of the pore dimensions could be described by Equation (4).

A different type of analysis has now provided this information (20). The dimension distributions $\rho(a)$ of independent spherical scatterers with uniform density and diameter a which produces each of the terms in the sum in Equation (3) can be calculated (19). After obtaining the constants in the sum in Equation (3) by least-squares fits of this equation to the scattering curve measured for Beulah lignite at the University of North Dakota, we used these constants to evaluate the sum of the pore-dimension distribution functions for uniform spheres that are obtained (19) from the terms in the sum in Equation (3). The sum of these pore-dimension distributions was very similar to the power-law distribution given by Equation (4). The fact that we could obtain almost the same power-law dimension distribution by two independent methods suggests that such a distribution may be a good approximation to the poredimensiom distribution in Beulah lignite, and, we expect, in other coals that have similar scattering curves.

By modifying and extending the results presented in Reference (19), we have developed a technique to calculate pore-dimension distributions from the scattering curves for lignites. Equation (5) was obtained under the assumption that the dimension distribution was given by Equation (4) for all values of the dimension a. Calculations (20) show that if, on the other hand, the scattered intensity can be approximated by Equation (5) for $h_1 \leqslant h \leqslant h_2$, and if the entire measured curve can be represented by the expression

$$I(h) = I_{oo} h^{-(7-\gamma)} + D_1 + D_2 h^{-4} \tag{6}$$

then Equation (4) gives the particle dimension distribution for dimensions in the interval $a_1 \leqslant a \leqslant a_2$, where
where

$$a_1 = 0.45/h_2$$

and
$$\tag{7}$$

$$a_2 = 3.5/h_1$$

The constants I_{oo}, D_1, and D_2 and the exponent $7-\gamma$ in Equation (6) can be evaluated by a least-squares fit.

Quantities calculated in our fits of Equation (6) to the scattering curves for lignites are listed in Table 3. The curves in Figures 3 and 4 were determined from the fits of Equation (6).

As can be seen from the curve for PSOC coal 93 in Figure 2, the scattering data from lignites can also be fitted quite well with Equation (3). The quality of the fits of the two equations is nearly the same. The fact that both equations can be employed in the fits is a result, we believe, of the fact that both Equations (3) and (6) correspond in this case to similar pore-dimension distributions.

Table 3
Quantities Obtained from the Fits of Equation (10)

Coal	Angles considered in the fit	$7-\gamma$	I_{oo}	D_1	D_2
PSOC 86	.003-.128 rad	2.8	27	5.2×10^3	0.58
PSOC 92	.003-.128	3.5	6.5	6.5	-0.14
PSOC 93	.003-.128	3.7	3.1	4.9	-0.13
PSOC 246	.003-.128	2.8	23	4.2	0.43
Beulah lignite (a)	.003-.128	2.4	28	4.2	0.56
Beulah lignite (b)	.0003-.100	3.44	4.7	1.5	0.0010

(a) Scattering curve measured at the University of Missouri
(b) Scattering curve measured at the University of North Dakota

The exponents $7-\gamma$ in Table 3 fall into two groups, with values less and greater than 3. The difference in exponents is compensated for by a change in the value of the constant D_2, so that the resulting fitted curves are nearly the same for all of the lignite samples, just as the measured scattering curves are almost identical. The data obtained from Beulah lignite at the University of North Dakota extend over a wider interval of scattering angles than the curve recorded for this coal at the University of Missouri and thus permit a better evaluation of $7-\gamma$ and the other constants in Equation (6) than is possible when the intensities are measured in a more resticted interval of scatterinmg angles. We therefore suggest that the exponent from the fit of Equation (6) to the scattering data measured at the University of North Dakota for Beulah lignite should give a good indication of the exponent describing the pore-dimension distribution in a typical lignite.

As can be seen from Table 3, when the exponent $7-\gamma > 3$, the constant D_2 is negative for the scattering curves for the PSOC coals. There is no reason (20) that these values cannot be negative.

Since the scattered intensity from lignites was proportional to a negative power of h for scattering angles from approximately 0.003 radians to 0.025 radians, we can conclude from Equation (7) that a_1 is at most about 4 Å and that a_2 must be at least 300 Å. As the data recorded for Beulah lignite at the University of North Dakota extend to angles as small as 0.0003 radians, this limit on a_2 may be at least ten times too small.

We would like to emphasize that our suggestion that Equation (4) approximates the pore-dimension distribution in lignites is only consistent with our scattering data, and other interpretations of the data cannot be excluded. We feel, however, that our data can be considered evidence that such a distribution is at least reasonable.

We will point out, nevertheless, that this distribution is valid only for $a_1 \leqslant a \leqslant a_2$, and that the values of $\rho(a)$ in this interval of a usually will not be sufficient to define the specific surface, which is proportional to

$$\int_0^\infty a^2 \, \rho(a) \, da$$

and the total pore volume, which requires evaluation of the integral

$$\int_0^\infty a^3 \rho(a) \, da$$

This result is a consequence of the fact that the specific surface and the pore volume may depend strongly on $\rho(a)$ for values of a outside the interval in which Equation (4) gives the pore-dimension distribution.

If our suggestion that Equation (4) approximates the pore-dimension distribution in lignites is correct, there is no unique way to classify the pores as macropores, transitional pores, and micropores, at least for pores smaller than about 3000 Å. Instead, there is a continuous distribution of the pore dimensions from about 4 Å to at least 3,000 Å. However, the fact that the pore classes may not be uniquely defined does not invalidate the alternative and essentially equivalent analysis of the data in which we employed Equation (3) to determine the x-ray specific surface.

Acknowledgments

We would like to express our gratitude to S. S. Pollack and H. K. Wagenfeld for their advice and assistance. We are very grateful to C. Philip Dolsen of Pennsylvania State University for supplying us samples of the PSOC coals and data on these coals. We also are pleased to acknowldge the help which Harold H. Schobert has given us, both by providing advice on a number of occasions and for supplying the Beulah lignite samples and the data on the analysis of these coals. Special thanks are extended to the Amoco Research Center, Naperville, Ill. for loaning the Bonse-Hart collimation system. Part of this work was supported by contracts from the U. S. Department of Energy and by the North Dakota Energy Research Center and the Associated Western Universities.

Literature Cited

1. Guinier, A.; Fournet, G.; Walker, C. B.; Yudowitch, K. L. "Small-Angle Scattering X-Rays"; Wiley: New York, 1955.
2. Durif, S. J. Chim. Physique 1963, 60, 816–24.
3. Kröger, C.; Mues, G. Brennstoff-Chemie 1961, 42, 77–84.
4. Spitzer, Z.; Ulický, L. Fuel 1978, 11, 621–625.
5. Lin, J. S.; Hendricks, R. W.; Harris, L. A.; Yust, C. S. J. Appl. Cryst. 1978, 11, 621–625.
6. Gonzales, J.; Torriani, I. L.; Luengo, C. A. J. Appl. Cryst. 1982, 15, 251–254.
7. Kalliat, M.; Kwak, C. Y.; Schmidt, P. W., in "New Approaches To Coal Chemistry"; (B. D. Blaustein, B. C. Bockrath, and S. Friedman, eds.), ACS Symposium Series No. 169, Amer. Chem. Soc.: Washington, 1981; pp. 3–22.
8. Setek, M. Thesis, Royal Melbourne Institute of Technology, Melbourne, Australia, 1983.
9. Wagenfeld, H. K.; Setek, M. Research to be published.
10. Ref. (1), pp. 3--4.
11. Kwak, C. Y.; Kalliat, M.; Schmidt, P. W. Research to be published. 1984. Kwak, C. Y. Ph. D. thesis, University of Missouri, 1982. (Copies available from University Microfilms, Ann Arbor, Mich.)
12. Kratky, O.; Skala, Z. Zeits. f. Elektrochem.-Ber. Bunsenges. phys. Chem. 1958, 62, 73–77.
13. Anderegg, J. W.; Beeman, W. W.; Shulman, S.; Kaesberg, P. J. Am. Chem. Soc. 1955, 77, 2929.
14. Bonse, U.; Hart, M. Appl. Phys. Lett. 1966, 7, 238–40. Bonse, U.; Hart, M. in "Small-Angle X-Ray Scattering", (H. Brumberger, ed.); Gordon and Breach: New York, London, 1967; pp. 121–130.
15. Dubinin, M. M.; in "Chemistry and Physics of Carbon" (Walker, P. L., Jr, ed.) Vol. II; Dekker: 1954; pp. 51–59.
16. Debye, P; Anderson, H. L., Jr.; Brumberger, H. J. Appl. Phys. 1957, 28, 679–683.
17. Gan H.; Nandi, S. P.; Walker, P. L., Jr. Fuel 1972, 51, 272–277.
18. Gregg, S. J.; Sing, K. S. W. "Adsorption, Surface Area, and

Porosity"; Academic Press: London and New York, 1967, pp. 211–217.
19. Schmidt, P. W. J. Appl. Cryst. 1982, 15, 567–569.
20. Bale, H. D., and Schmidt, P. W. unpublished research.

RECEIVED March 5, 1984

Determination of the Microstructure of Wet and Dry Brown Coal by Small-Angle X-Ray Scattering

M. SETEK[1], I. K. SNOOK[2], and H. K. WAGENFELD[2]

[1]Bureau of Meteorology, Melbourne, Victoria 3000, Australia
[2]Department of Applied Physics, Royal Melbourne Institute of Technology, 124 LaTrobe Street, Melbourne, Victoria 3000, Australia

The identification of the microstructure of coal is a necessary prerequisite for its efficient utilization (1). Some of the parameters that characterize the microstructure are the porosity, the micropore volume and surface area. In Australia, these parameters are routinely measured on dry Victorian brown coal at the Herman Laboratory (State Electricity Commission of Victoria), Richmond, Victoria, by means of the well-established technique of gas adsorption (GA). But this method works only on coals which have been dried. An alternative method, which can be used for wet coals too, consists of an analysis of the intensity distribution of X-rays which have been scattered in coal at small scattering angles. This method is known as Small Angle X-ray Scattering (SAXS). It was firstly applied by Krishnamurti (2) in the early 1930s to colloidal solutions and other appropriate systems. However, a systematic investigation of this method was first started in the late 1930s by Guinier (3,4). Today, SAXS is used widely for the investigations of porous materials and other systems and has been applied to coal (5-12).

The application of SAXS to an investigation of the microstructure of coal seems to be very attractive in the first instance. However, the shape of the pores is unknown and it is therefore not obvious how to characterize the pore structure of such an heterogeneous material as brown coal. Even defining what we mean by structure is difficult for a system with such diverse constituents as water, carbon, minerals, plant matter, pollen, etc. and whose composition may even vary within a seam. According to a model by Hirsch and Cartz (14, 15), the characteristic of low-rank coals (up to 85% carbon) is the 'open structure' consisting of randomly-oriented aromatic layers, each containing 2 or 3 fused benzene rings. These layers are interconnected by aliphatic crosslinks resulting in a structure that is highly porous. With increasing rank the pores tend to disappear gradually and the result is the so-called 'liquid structure' characteristic of bituminous coal (85-91% carbon). The aromatic layers begin to display some order, and there are less crosslinks. During the final stages of coalification a new pore system may be formed. This is the 'anthracitic structure' and is characteristic of high rank coal

0097-6156/84/0264-0095$06.00/0
© 1984 American Chemical Society

above 91% carbon. In this structure the aromatic layers are highly oriented and crosslinks are almost absent resulting in an oriented porous system.

Victorian brown coals are thought to be largely amorphous, containing aromatic layers of single substituted benzene rings crosslinked by aliphatic chains to form a three dimensional structure. Their carbon content is quite low, varying from 60 to 70%. One would therefore expect its porous system to be somewhat like that of an 'open structure' having micropores which are randomly-oriented. In this preliminary study two samples of Yallourn ream coal were taken from the Yallourn open cut mine in the Latrobe Valley, Victoria, Australia. The samples, a pale and a medium dark lithotype, are representative of the extremes in coal types found in the Yallourn ream.

The scattering at small angles (less than 5°) is caused by the pores of the coal. Many materials produce similar scattering and in most cases it is caused by particles with short range order embedded in a medium of different electronic density. However, dry brown coal produces a very intense scattering very near to the primary beam due to the presence of microvoids, not particles. This is not obvious, because by Babinet's optical principle of reciprocity (4), voids and particles produce the same diffraction patterns, unless the particles have a high X-ray absorption coefficient. In order to find out whether the pores are filled or not filled, one has to carry out adsorption measurements. Since the adsorption of gas can only take place in voids, a reasonable agreement of the data for the pore structure determined by means of GA and SAXS, proves that the scattering is due to voids in dry coal. Kalliat et al (9) find in their investigations of coal the presence of three pore size classes: large pores, with dimensions of at least 300 Å, intermediate pores, with average dimensions of the order of 100 to 200 Å and very small pores, which have dimensions not greater than about 30 Å. These pores are referred to as macropores, transition pores and micropores, respectively. This classification is roughly consistent with ours. The total pore volume, V_p, is measured in cm^3/g. The macropores contribute mainly to this quantity. The surface area, S_p, is measured in m^2/g. It will be seen later that the micropores contribute mainly to the surface area. It is however obvious that S_p depends critically on the shapes assumed for the micropores. The shapes of the micropores are unknown and one has to make assumptions, i.e. use simple models of the shapes of micropores which match the experimental data. But there is obviously no unique solution to this problem. Careful investigations by means of electron transmission microscopy might be helpful in order to find some information about micropore shapes.

Experimental

Apparatus Description. An accurate small angle scattering experiment requires a camera with a high angular resolution. Kratky (16) designed a camera which meets this requirement and a Kratky Camera was therefore used in our experiments. This instrument has a slit system designed such that it allows measurement of the scattered intensity very close to the primary beam. Its angular resolution is better than 20 seconds of arc. Figure 1 shows the

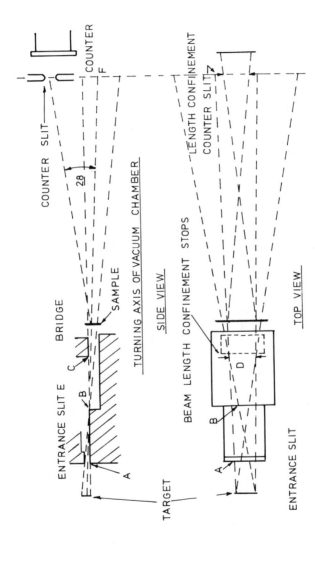

Figure 1. Collimation system of Kratky camera showing the beam geometry.

collimation system of the Kratky Camera. The X-ray beam enters through entrance slit E, its width in the vertical direction being defined by the optically polished collimating surfaces A, B and C. The length of the primary beam in the horizontal direction is determined by the length of the line shaped focus (target) in the X-ray tube and the length of the confinement stops at D. The purpose of the bridge C is to reduce the parasitic slit edge scattering introduced at the collimating surfaces A and B, such that the detector slit can be positioned to within 100 μm or less of the center of gravity of the primary beam.

In all experiments Cu K_α (λ = 1.54 Å) radiation was used. The monochromatisation of the X-ray beam was achieved by using a nickel filter together with a pulse height analyzer in order to exclude the Cu K_β radiation. Such a method reduces all radiation apart from the K_α line to below 1%.

The detector is a Siemens Ar-Me proportional flow counter which has high detection efficiency for Cu K_α radiation.

The mechanical stability of the Kratky Camera may be expressed as a change in the position of the center of gravity of the primary beam. This is less than 5 seconds of arc over a period of a few days.

Sample Preparation. Several coal samples (light and dark lithotypes) were examined in their dry and wet stages. Different preparaton techniques and sample holders had to be used for the dry and wet samples. In order to achieve sample homogeneity in the case of dry coal, the coal was at first crushed to a fine powder and passed through a 200 μm sieve. It was then either air or vacuum dried and packed into a thin wall glass capillary of dimensions less than 1 mm in diameter and sealed. The capillaries are manufactured from a special silica glass almost free from impurities such that the contribution to the scattering at low angles is minimal. As wet samples cannot be treated in the same way as dried samples, i.e. it is impossible to get a wet sample into a glass capillary, we had to produce a different sample holder so that the wet samples can remain in its wet moist state. When extracted from underground, it contains approximately 60 to 70% of moisture by weight. The sample is crushed as quickly as possible and packed into a special sample cell with low absorbing 1 μm mica windows. Glass capillaries are impractical for wet samples since it is very difficult to pack them with sufficient speed to prevent the moisture from escaping.

Absolute Intensity Measurement. The radius of gyration is the only precise parameter which can be determined from the scattering curve without invoking further auxiliary hypothesis and further information can be determined from the scattering curve, provided the ratio of the scattered intensity to that of the primary beam intensity is known (the intensity of the primary beam is usually measured after it passes through the sample in order to avoid absorption measurements). This ratio is known as the absolute intensity in small angle scattering experiments (17). The knowledge of this ratio can be used, for example, in the determination of the volume of the scattering particle. This is very useful auxiliary information since knowing the particle volume and its radius of gyration, its shape can also be determined.

To calculate the absolute intensity it is necessary to know the strength of the primary beam. This can be achieved by using a specially prepared standard sample that scatters X-rays in a known way such that the absolute intensity scattered by any sample is determined by a simple comparison calculation. It must be sufficiently insensitive to the action of X-rays ([18]) as well as being homogeneous throughout. The standard sample used here is Lupolen (a polyethelene platelet) supplied with the Kratky Camera.

The scattering power of this sample is determined by what is known as the rotator method ([19]). This measurement is done using a modified Kratky Camera that incorporates a rotating disc with equally spaced small holes near its perimeter. The disc is fixed at the registration plane such that the holes pass through the primary beam while it is rotating. The rotation speed is such that only few X-ray quanta enter the detector during the time a particular hole is travelling through the primary beam. The number of X-ray quanta entering the detector per unit time is thus well below its linear response. Knowing the diameter of the holes and the number of revolutions per minute it is possible to calculate the strength of the transmitted beam (P_c) in counts per unit time per one centimeter of its length in the horizontal direction.

The scattering curve of the standard sample (Lupolen) is shown in Figure 2. It contains one broad maximum followed by a uniformly and rapidly decreasing region. Choosing the scattering angle 2θ in approximately the middle of this uniform region (this angle is usually chosen to correspond to a Bragg spacing of 150 Å), where the collimation error is almost negligible ([17]) the scattered intensity $I_{2\theta}$ at that angle in counts per unit time, is measured. With the Kratky slit system the scattered intensity decreases as $1/R$. It therefore follows that $R\, I_{2\theta}/P_c$ is the constant which we call $K(K = R\, I_{2\theta}/P_c)$. The constant K together with the standard sample transmission factor T_c is already given. To determine the strength of the primary beam after it passes through the sample (P_s) it remains to measure the sample transmission factor T_s. Using the fact that $P_s/P_c = T_s/T_c$ it follows that

$$P_s = K\, I_{2\theta}\, \frac{T_s}{T_c}\, R \qquad (1)$$

where P_s is the number of counts per unit time for a primary beam of one centimeter length. For our Lupolen sample

$$P_s = 70.4\, I_{2\theta}\, T_s\, R \qquad (2)$$

The transmission factor (T_s) was measured by using the standard sample as an auxiliary scatterer. This is shown schematically in Figure 3 below. The coal sample is placed in position A and the intensity scattered by the standard sample is recorded. The coal sample is then taken out and again the scattered intensity is recorded. This is then repeated at least eight times. The transmission of the glass capillary or a sample cell is determined by the same method.

Finally, to calculate the transmitted intensity of the primary beam (after it has passed through the sample) the quantity P_s is divided by the effective vertical width (W) of the primary beam as

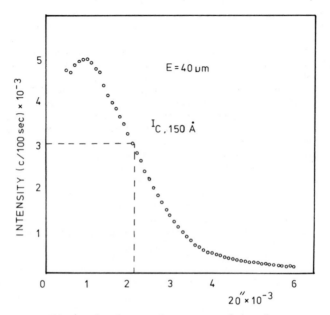

Figure 2. Scattering curve of lupolen.

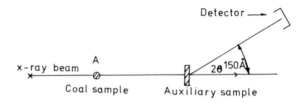

Figure 3. Experimental setup for determining T_s.

seen by the detector. The quantity W is similar to the width of a Bragg reflection at half its maximum and is derived from the vertical beam profile. For a 40 µm entrance slit, W = 98 µm.

Analysis of Experimental Data

Samples of light and dark lithotype coals in their dry and bed-moist states have been investigated. Their computer unsmeared scattering curves are shown in Figure 4. They all possess very similar shape characteristics, such as strong concave character, steep negative slope at lower angles, and at high scattering angles a long slowly varying tail. This type of scattering curve is assumed to be due to the presence of a dilute system of micropores which vary widely in size but are approximately identical in shape. For simplificaton we assume that the pore size distribution is a Maxwellian distribution. This assumption is certainly over-simplified for such a complex system as brown coal.

Mean Radius of Gyration. The mean radius of gyration, R, of the pores can be obtained from the so-called Guinier plot (20):

$$\ln I(h) = -\frac{h^2 R^2}{3} + \text{constant} \tag{3}$$

where

$$h = 4\pi \sin\theta/\lambda \tag{4}$$

λ is the X-ray wavelength and θ half the scattering angle (Bragg angle). Thus by plotting the logarithm of intensity versus h^2 the parameter R can be calculated from the slope. The radii R for the four different coal samples are listed in Table I. The obvious result that the coal shrinks during the process of drying is reflected in the fact that the radius of gyration is smaller in the dry state relative to the wet state.

Micropore Volume. A calculation of the specific micropore volume of a sample requires the knowledge of its microporosity, given by

$$P = 100 \ (1-c) \ \% \tag{5}$$

where c is the fraction of the sample occupied by solid, and this can be obtained from the intensity distribution (21). The micropore volume is then given by

$$V_p = \frac{P}{100\rho} \ cm^3/g \tag{6}$$

ρ is the apparent sample density, that is, mass of sample per unit volume which includes the solid medium and all pore space. The experimentally determined values for V_p are given in Table I.

Number of Micropores, Average Micropore Volume. Another important quantity that can be calculated from the knowledge of absolute intensity is the number of micropores present in the sample. It may

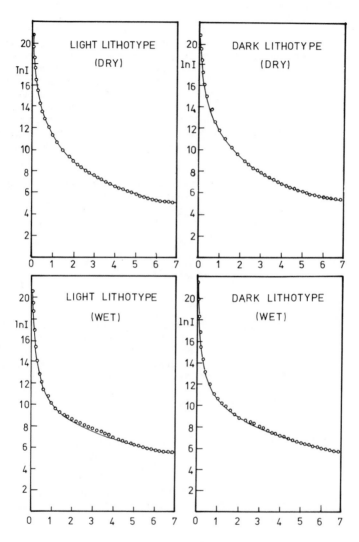

Figure 4. Collimation-corrected scattering curves of coal samples (abscissa, 2Θ " $\times 10^{-3}$).

be shown that the scattered intensity at zero angle is given by

$$I(0) = 1.867 \times 10^{-28} \, I_o \, \bar{N}n^2 \tag{7}$$

where $I(0)$ was computed for each sample by the numerical method (4). Knowing I_o it is seen that Equation 7 gives the absolute values of $\bar{N}n^2$, where $\bar{N}$ represents the number of micropores in the portion of sample irradiated by X-rays. The quantity $\bar{N}n$ gives the difference between the number of electrons contained in micropores and the irradiated volume of coal medium. It is calculated by the following formula

$$\bar{N}n = N_A \, m \sum_i F_i \, A_i/W_i$$

where F_i is the fraction of ith element in the sample, A_i and W_i are the atomic and mass number of the ith element respectively, and m is the mass of the irradiated sample volume V_s given by $\Delta\rho \, V_s$. N_A Avogadro's number. Since $A_i/W_i \sim 2$, the above equation may be rewritten as

$$\bar{N}n = \frac{V_s \, N_A}{2} \, \Delta\rho \tag{8}$$

where $\Delta\rho$ is the difference in specific densities between the pore and coal media. It is calculated by

$$\Delta\rho = \rho_{He} - F_1\rho_{water} - F_2\rho_{air} \tag{9}$$

ρ_{He} is the helium density of the sample measured by helium displacement. F_1 and F_2 correspond to the fractions of water and air content respectively in the micropore system. These can be calculated from the moisture content of raw coal (bed moist coal) and the weight loss upon air drying. Combining Equation 7 and 8 and dividing by $\rho_{Hg} \, V_s$ gives the number of micropores per gram of coal.

$$\bar{N} = 1.693 \times 10^{19} \, \Delta\rho \, V_s \, I_o/\rho_{Hg} \, I(0) \tag{10}$$

The average volume of a sngle micropore $(\bar{v})$ is then given by

$$\bar{v} = 10^{24} \, \frac{V_p}{\bar{N}} \, \text{in Å}^3 \tag{11}$$

Results are shown in Table I.

Pore Size Distribution. The scattering diagrams of the coal samples shows a very pronounced curvature followed by a long linear part with a slight slope. It is therefore assumed that the micropore system of brown coal consists of a mixture of different size pores. The shape of the curves indicates the presence of large micropores with a large proportion of small micropores, the larger pores giving rise to the extremely steep initial part of the scattering curve while the small pores give rise to the long slightly inclined tail.

Table 1

Small Angle X-ray Scattering (SAXS) and Gas Adsorption (GA) Data for Brown Coal (See Reference (12)).

SAMPLE	DARK LITHOTYPE			LIGHT LITHOTYPE		
METHOD	GA	SAXS		GA	SAXS	
STATE OF SAMPLE	DRY	DRY	WET	DRY	DRY	WET
MICROPORE VOLUME - V_p (cm^3/g)	0.079	0.095	0.100	0.058	0.060	0.110
SPECIFIC SURFACE AREA - S_p (m^2/g)	298	S_o 91 S_p 1130	S_o 34 S_p 213	216	S_o 72 S_p 828	S_o 42 S_p 393
RADIUS OF GYRATION - R (Å)		830	1580		720	1360
SPECIFIC NUMBER OF MICROPORES $\bar{v}$ (Å^3)		1.1×10^7	3.0×10^8		8.6×10^6	8.5×10^7
AVERAGE VOLUME OF SINGLE MICROPORES $\bar{N}$ (cm^3)		8.9×10^{15}	5.0×10^{14}		6.9×10^{15}	1.3×10^{15}

Specific Surface. A first possibility is to assume a certain shape
of the pores and carry out a summation over all pores. If one for
instance assumes that the pores are ellipsoids and chooses a Maxwell
distribution function then one obtains for prolate ellipsoids
(Figure 5):

$$S_p = 10^{-19} \ 2\pi \ \bar{N} \ r_o^2 \ \frac{\gamma}{(2 + \gamma^2)} \ \frac{\Gamma(\frac{n+3}{2})}{\Gamma(\frac{n+1}{2})} \tag{12}$$

and for oblate ellipsoids (Figure 6)

$$S_o = 10^{-19} \ \frac{\pi \ \bar{N} \ r_o^2}{(2 + \gamma^2)} \ \frac{\Gamma(\frac{n+3}{2})}{\Gamma(\frac{n+1}{2})} \tag{13}$$

The axes of the ellipsoids are: a, a, and aγ . r_o is the
classical electron radius. Numerical values obtained for the
surface areas are given in Table I. It is obvious that the values
for the surface area depends critically on the shape of the pores.
A second method is to use the so-called Porod limit (22-24).
Porod investigates the tail of the intensity curve in order to get
the surface area. One has to check this method whether the product
h^4I versus h^4 is a straight line. Unfortunately the plot of h^4I
versus h^4 shows only a very brief linear region with a positive
slope that seems to curve upwards with increasing 2Θ. This deviaton
from Porod's law is revealed more so in the plots of h^4I - h^3, h^4I
- h^2, h^4I - h, the last curve having the longest linear region
suggesting the following asymptotical behaviour

$$I(h) = a \ h^{-3} + b \ h^{-4}$$

The specific surface area determined from the Porod plot is
approximately 10 m^2/g. This is a very low value and can be
attributed to the absence of the Porod limit.

Discussion of Results. The term 'pore' has been and will be used
frequently. It is therefore essential to ascertain its precise
meaning. Pores are small cavities in a material which may range
considerably in size and shape. A pore can be so large that its
interior walls can be considered smooth relative to the atoms
composing them. Such pores are called macropores and may be up to
several microns across. On the other end of the size scale a pore
can be formed within a small group of atoms and is then considered
as any space free of atomic bonds. Such spaces are accessible by
small atoms or molecules and are called micropores.
Results from SAXS analysis and GA (using CO_2 gas) for the same
coal samples are shown below in Table I. From the results of SAXS
analysis it may be concluded that the investigated Victorian brown
coal possesses an extensive micropore system containing between 10^{14}
to 10^{16} pores per gram. Additionally, according to the shape
hypothesis these pores may be either slit-like (thin discs with
large diameters) or filament-like (long narrow cylinders) and are

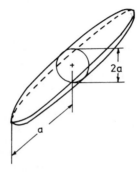

Figure 5.   Prolate ellipsoids, where d = 2a, and L = 2aγ.

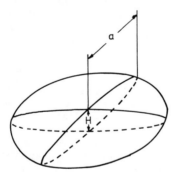

Figure 6.   Oblate ellipsoids, where D = 2a, and H = 2aγ.

present in a wide size distribution. In dried coal the slit-like pores appear to be very narrow (less than 5Å). It is difficult to consider this dimension as a separating distance between two surfaces, since for most but the smallest adsorbate species (such as helium) these pores cannot be detected. Only a single helium molecule can be accommodated between two atomic layers separated by 3Å, hence the two surfaces would be 'seen' as one. For coal samples in bed moist condition this dimension increases several times, which indicates shrinkage occurs in the microporous structure of brown coal upon drying. This behaviour is exhibited by the increase in the micropore volume and the pore size parameters.

Of the results obtained comparison of SAXS with GA can be best made by comparing the micropore volumes, since it is this property of the coal's physical structure which is directly observed. Results for the air dried coal samples show reasonable agreement between the two methods. However, this agreement is not apparent in the surface area results. In both methods surface area is calculated from the measured micropore volume and in each instance certain assumptions must be made. As an example in GA method the hypothesis is dependent on the type of molecular packing in the adsorbed monolayer and the area occupied by each adsorbate molecule. In SAXS method the geometrical form of the micropores must be assumed in order to compute their surface area. It is a well-established fact that if discrepancies do occur in comparison of specific areas between the two methods, it is the small angle values that are almost always larger (25). Many investigators have interpreted this regular disparity as due to the existence of closed pores or pores inaccessible to the adsorbing molecule. As an example, for those micropores where the minimum separation distance is only 3Å, full penetration of such pores could not be achieved and hence less surface is detected by the adsorbate gas.

None of the surface areas calculated for prolate or oblate ellipsoid-like pores shows a good agreement with the GA values. The values for the surface area obtained by GA or SAXS remains uncertain. The values given by Kalliat and co-workers (10) refer most likely only due to the large pores.

This investigation will be continued. Different coal specimens will be investigated under different physical conditions in particular different moisture contents.

Acknowledgments

We are very grateful to Messrs Kiss and Stacy, State Electricity Commission of Victoria, Herman Laboratories, Richmond, Victoria, for giving us their Gas Adsorption data and for lending us the Kratky Camera. In addition, we thank the Brown Coal Council, Victoria, for their continued financial support and, the National Energy Research, Development and Demonstration Council for the grant to purchase an updated Kratky Camera.

Literature Cited

1. Walker, P.L. Phil. Trans. R. Soc. Lond. A. 1981, 300, 65.
2. Krishnamurti, P. Indian J. Phys. 1930, 5, 473.
3. Guinier, A. Compt. rend, 1938, 206, 1376.
4. Guinier, A. and Fournet, G. "Small-Angle Scattering of X-rays" 1955, J. Wiley, New York.
5. Riley, D.P. Brit. Coal Utilization Research Assoc. Conf. 1944, 232, London.
6. Durif, S. J. Chim. Physique, 1963, 60, 816.
7. Kroger, C., Murs, G. Brennstoff-Chemie 1961, 42, 77.
8. Spitzer, Z., Ulicky, L. Fuel, 1976, 55, 21.
9. Lin, J.S., Hendricks, R.W., Harris, L.A., Yust, C.S. J. Appl. Cryst. 1978, 11, 621.
10. Kalliat, M., Kwak, C.Y., Schmidt, P.W. New Approaches to Coal Chemistry 1981, 169, 1.
11. Wagenfeld, H.K., Setek, M., Kiss, L. and Stacy, W.O. Proceedings of the International Conference on Coal Science 1981, 869-874, Dusseldorf, FRG.
12. Setek, M., Wagenfeld, H.K., Stacy, W.O. and Kiss, L. Fuel 1983, 62, No. 4, 480.
13. Setek, M. MSc Thesis, Royal Melbourne Institute of Technology, 1983, Melbourne, Australia.
14. Hirsch, P.B. Proc. Roy. Soc. A. 1954, 226, 143.
15. Cartz, L., Hirsch, P.B. Phil. Trans. R. Soc. Lond. A. 1960, 252, 557.
16. Kratky, O., Skala, Z. Zeits. Electrochem.-Ber.Bunsenges. Phys. Chemie, 1958, 62, 73-77.
17. Brumberger Small Angle X-ray Scattering, Gordon and Breach, Science Publishers, New York, 1967.
18. Kratky, O., Pilz, I., and Schmitz, P.J. J. Coll. Interface Sci. 1966, 21, 24.
19. Warren, B.E., Z. Krist. 1959, 112, 255.
20. Guinier, A. Ann. Phys. 1939, 12, 161.
21. Reference (1), page 157, Equation (33).
22. Porod, G. Kolloid Z. (1952), 124, 83.
23. Porod, G. Kolloid Z. (1952), 125, 51.
24. Porod, G. Kolloid Z. (1952), 125, 109.
25. Janosi, A. and Stoeckli, H.F. Carbon, 1979, 17, 465.

RECEIVED July 11, 1984

Chemical Variation as a Function of Lithotype and Depth in Victorian Brown Coal

R. B. JOHNS, A. L. CHAFFEE[1], and T. V. VERHEYEN[2]

Department of Organic Chemistry, University of Melbourne, Parkville, Victoria 3052, Australia

Victorian brown coal occurs in five major lithotypes distinguishable by color index and petrography. Advantage has been taken of a rare 100 m continuous core to compare and contrast chemical variations occurring as a function of lithotype classification. For many parameters there is a much greater contrast between the different lithotypes than there is across the depth profile of (nearly) identical lithotypes. Molecular parameters, such as the distributions of hydrocarbons, fatty acids, triterpenoids and pertrifluoroacetic acid oxidation products, together with gross structural parameters derived from IR and ^{13}C-NMR spectroscopic data, Rock-Eval and elemental analyses and the yields of specific extractable fractions are compared. Correlations with paleobotanical assessments of the environments of deposition of the lithotypes are suggested.

Victorian brown coal occurs in randomly sequenced stratified layers, known as lithotypes, which are distinguishable by their air-dried color, maceral composition and many physical and chemical properties. The five major lithotypes can be related to generalized paleo-environments of deposition and are known to influence many coal utilization parameters (1-3). Across the thick coal intervals which occur (up to 300 m) there is a small increase in coal rank observable via parameters such as percent carbon and percentage volatiles (1). The purpose of this paper is to contrast the variation in the organic chemical nature of the coal occurring as a function of depth for a series of (nearly) identical lithotypes with that occurring as a function of lithotype. We have employed techniques giving both precise molecular level information (e.g., the distribution of extra-ctable lipid classes, composition of Deno oxidation (4) products) and average structural characterization of the whole coal (e.g., para-meters derived from IR, solid state ^{13}C-NMR, etc.). All of these techniques have been able to provide much deeper insights into the

[1]Current address: Commonwealth Scientific and Industrial Research Organisation, Division of Energy Chemistry, PMB 7, Sutherland, N.S.W. 2252, Australia
[2]Current address: Victorian Brown Coal Council, G.I.A.E., Switchback Road, Churchill, Victoria 3842, Australia

0097-6156/84/0264-0109$07.25/0
© 1984 American Chemical Society

varying nature of the coal than the conventional elemental and functional group type analyses which have been carried out extensively
in the past.

Materials and Methods

Coal samples were taken from a single bore core (LY 1276) from the
Flynn field in the Loy Yang region of the Latrobe Valley, Victoria,
Australia. The core consists of > 100 m of continuous coal, but
penetrates two coal seams, viz. Morwell 1A and Morwell 1B. These
seams range from late Oligocene to Miocene in age. It is estimated
that deposition of the 100 m of coal would have taken approximately
1 million years. Samples were chosen by visual examination of the
air-dried color of the coal. Lithotype classifications on this basis
are usually, but not always, correct; it was therefore found necessary to modify some of the initial classifications after a consideration of all the available petrographic and chemical data. This
correction has not been incorporated in previous publications from
our group (5).

Experimental Techniques:

(A) Solvent Extraction - freeze dried lithotype samples were subjected to $CHCl_3$/MeOH (2:1; v/v) and toluene/MeOH (3:1; v/v) extraction using both Soxhlet and sonication apparatus. Solvent extractable humic acids (SE.HA) were isolated via sonication of the total
solvent extractable material with 0.5 N NaOH + 1% $Na_4P_2O_7$ and subsequent acidification to pH 2 (7). Lipid materials including hydrocarbons and fatty acids were isolated via saponification in $MeOH/H_2O$
(5:1; v/v) with KOH (8) followed by acid/base extraction. The neutral and acidic fractions were further separated by thin layer
chromatography (Silica GF_{254}) using n-heptane/toluene (95:5; v/v) and
n-heptane/diethyl ether/methanol (80:20:2; v/v/v) respectively.
Polycyclic aromatic hydrocarbons (PAH's) were isolated by high performance liquid chromatography using Cyanobonded Spherisorb silica
(5 μ) columns with n-heptane as eluent and UV (254 or 310 nm)
detection (8).
     Gas chromatography and gas chromatography/mass spectrometry were
employed to identify and quantitate individual molecular components.
Both 25 and 50 meter glass support coated open tubular (SCOT) and
fused silica wall coated open tubular (WCOT) capillary columns (SE.30,
BP.1 and BP.5 phases) were used with H_2 as a carrier gas and F.I.D.
detection. Acidic components were derivitized (BF_3/methanol) to
their methyl esters and hydroxyl groups to their silyl ethers (N,O-
bis-(trimethylsilyl)trifluoroacetamide) in order to improve chromatographic separation. Carbon Preference Indices (CPI) were calculated using the equation -

$$CPI = \frac{1}{2} \left[\frac{\Sigma \text{ odd } C_{15}-C_{35}}{\Sigma \text{even } C_{16}-C_{36}} + \frac{\Sigma \text{ odd } C_{17}-C_{37}}{\Sigma \text{even } C_{16}-C_{36}} \right]$$

(B) Spectroscopic Analyses. (i) Infrared spectra were recorded
from KBr discs in absorption mode using the techniques and equipment
outlined in reference (7). Integrated absorption coefficients
(K cm.mg^{-1}) = absorbance unit x wavenumber x mg^{-1} x cm^2 are explained
in (9). (ii) Solid State ^{13}C-NMR spectra were obtained using cross

polarization, magic angle spinning techniques at 15.1 MHz employing
the instrument and conditions described in (10).

(C)   Rock-Eval Pyrolysis - was performed on the samples using the
equipment and techniques outlined in (11). The technique employs
automated equipment providing controlled heating (25°C/min) of the
sample under N_2 from ambient to 550°C with subsequent detection and
quantitation of evolved hydrocarbons and CO_2.

(D)   Bond equivalence data are expressed via a ternary diagram using
the method described in reference (12). Briefly, the bond equivalence percentages are calculated by dividing the Wt% C by 12/4, Wt% 0
by 16/2, Wt% H by 1/1 and renormalizing. Sulphur and nitrogen are
accounted for by incorporating their values within the oxygen
calculation.

(E)   Pertrifluoroacetic Acid Oxidation was performed on the coals
using the methods described in reference (4,13). Particular care was
taken to extract the glassware used for the oxidation with CH_2Cl_2 to
ensure removal of long chain (> C_{16}) material which is insoluble in
the aqueous product mixture. Aliphatic protons present in the product mixture were quantitated using 200 MHz proton magnetic resonance
(^{1}H NMR) spectroscopy. The products were subsequently fractionated
into neutral, acidic and polar acidic moieties using solvent extraction and XAD-8 resin adsorption techniques (13).

Characterization of Whole Coals

Before considering the results of other more sophisticated analyses
it is useful to demonstrate the divergent chemical nature of the coal
lithotypes by a more classical approach. When elemental analytical
data are expressed in bond equivalence form it can be seen (Figure 1)
that there is some inherent spread in the plotted points for the
suite of *light* lithotype samples. No consistent trends are observed
within the light lithotypes with depth. The different lithotypes
are, however, readily distinguishable due to the much greater varia-
nce in their analytical figures. The distinguishing features are
especially exhibited at the extremes of the lithotype classification
as an increase in the number of bonds associated with carbon at the
expense of hydrogen for the *dark* lithotype and visa versa for the
pale lithotype. For the *light* samples the bond equivalence data are
most consistent for carbon, with the mean values for hydrogen and
oxygen exhibiting higher proportional standard deviations. The bond
equivalence data implies that the difference between the lithotypes
resides in the type of chemical structure(s) that the elements form.
     There are only a few solid-state techniques capable of providing
structural information on "whole" coals. The cross polarization,
magic angle spinning (CP-MAS) ^{13}C-NMR spectra of Victorian brown
coals exhibit significant fine structure including phenolic and
carboxylic resonance envelopes which are consistent with their low
aromaticity and rank. The proportion of aromatic carbon atoms, f(a),
in the coal can be determined from the spectra and the variation in
this parameter across the sample suite is illustrated in Figure 2.
While there is some spread in the f(a) values amongst the *light*

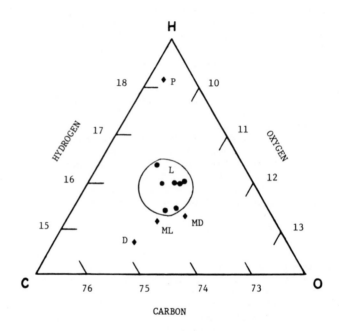

Figure 1   Ternary % bond equivalence diagram for whole coal
            samples.
            - *Light* lithotype samples; - other lithotypes as
            labelled:
            P - *pale*, ML - *medium light*, MD - *medium dark*, D - dark.

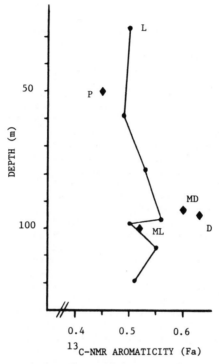

Figure 2   Aromaticity versus depth.
Data for the *light* lithotype samples are joined to
illustrate the spread of values.

sample (χ = 0.52, σ = 0.03) there is no obvious trend with depth.
The variation as a function of lithotype (sampled over a narrow depth
range) transcends the inherent spread in the *light* samples and there
is an obvious increase in aromaticity for the progression from paler
to darker lithotypes which is independant of depth. Only the *medium-
light* sample cannot be adequately distinguished within the sample
suite by this technique. In addition to the higher aromaticities the
spectra of the *medium-dark* and *dark* lithotypes exhibit prominent
phenolic and methoxyl carbon absorptions similar to the spectra of
degraded lignins. Hence the variation in f(a) as a function of
lithotype is considered to be predominantly the result of variation
in the original organic input and in the relative preservation of
various biopolymeric materials within the corresponding depositional
paleoenvironments. Although aromaticity is known to correlate with
coal rank over a wider regime, there is no suggestion of a regular
increase in f(a) over the depth interval examined. The small
variation in f(a) for the several *light* samples is probably a
reflection of minor changes in the depositional paleoenvironments.

Absorption mode IR spectroscopy as a technique, provides greater
sensitivity for the observation of specific functional groups. The
variation in the absorption coefficients K_{2920} (aliphatic C-H
stretch) and K_{1710} (carbonyl, carboxyl stretch) across the suite of
samples is illustrated in Figure 3. It is again obvious that the
variation as a function of lithotype exceeds the spread inherent in
the set of *light* samples. The proportional spread for the *light*
samples is significantly more intense for K_{2920} (χ = 14.1 cm mg^{-1},
σ = 2.9) than for K_{1710} (χ = 36.2, σ = 2.2), but in neither case is
there any definitive trend with depth. A strong negative correlation
(r = -0.95) exists between K_{1710} and f(a) for all samples examined.
The correlation between f(a) and K_{2920} is significantly lower
(r = -0.84) as would be expected from the higher proportional spread
in the K_{2920} values. The structural implications of these relationships are not entirely clear, but the IR data provide a further
indication of the significant effect which depositional paleoenvironment has upon the coal structure(s) which are ultimately preserved.

Rock-Eval analysis is a form of temperature programmed pyrolysis
often applied to petroleum source rocks in which the level of evolved
carbon dioxide and hydrocarbon-like material are quantitatively
measured via oxygen (OI) and hydrogen (HI) indices. The temperature
at which the maximum amount of volatile material is evolved (T_{max}) is
also recorded, and this can be related to the rank of the organic
substrate. The average T_{max} values for the samples under consideration was observed to be $397^{0}C$ (σ = 2.16), a value which corresponds
to a vitrinite reflectance of $\sim$0.30. No significant regular variation in T_{max} was observed with either depth or lithotype and this
method, also, does not provide any indication of a slight increase in
rank over the 100 m coal interval. A plot of HI versus OI (Figure 4),
however, provides a further method for distinguishing the lithotypes.
Although there is a significant spread in the HI values for the *light*
samples, the OI data enable better discrimination between the different lithotypes. This observation supports the NMR and IR data in
suggesting that lithotype is a coal property especially influenced by
the nature and concentration of oxygen containing species. HI is a
measure of the hydrocarbon production potential and correlates

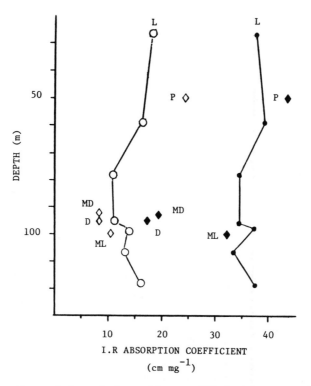

<u>Figure 3</u>   Infrared absorption coefficients versus depth.
Closed symbols refer to K_{1710}; open symbols refer to
K_{2920}.

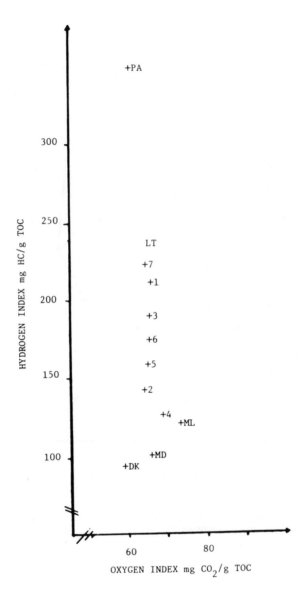

Figure 4   Hydrogen index (HI, mg HC/g TOC) versus
                 oxygen index (OI, mg CO_2/g TOC) from Rock-Eval analysis.

strongly (r = +0.94) with the percentage of extractable material (see later); it is a parameter which has substantial significance for coal-to-oil conversion. It can be seen that the lighter lithotypes and especially the *pale* sample provide greater potential in this respect.

The pertrifluoroacetic acid oxidation tecimique (Deno oxidation) has been applied to the lithotype and depth profile sample to investigate the structure, chain length distribution and concentration of their aliphatic components at the molecular level. Quantitative ^{1}H NMR spectroscopic analysis of the aliphatic protons present in the total oxidation product mixtures coupled with elemental analysis of the parent coals, enabled the percentage of initial coal hydrogen content present as aliphatic hydrogen in their oxidation products to be calculated. This data is plotted in Figure 5. The data exhibits no clear trend with depth (Figure 5) although there is a tendency for the light lithotypes to report a higher weight percent yield of aliphatic hydrogen in comparison to the dark lithotypes. Despite the oxidation technique providing secondary structural information, the plot is consistent with direct ^{13}C-NMR, IR and Rock-Eval analyses in reflecting no regular variation in aliphatic content with depth.

The lithotype profile was investigated in greater detail with the product composition of the different brown coals being reported in Tables 1 and 2. A distinct decrease in the total concentration of detectable oxidation products occurs with darker lithotypes (Table 1). This result is consistent with increasing aromaticity (Figure 2) and the preferential attack on aromatic structures by the pertrifluoroacetic acid reagent. The total destruction of the tertiary structure within the brown coal lithotypes is evidenced by their low yield of insoluble products (Residue) which is primarily composed of mineral matter.

Acetic, succinic and malonic acids (Table 1) are reported (4) to derive from Ar-CH$_3$, Ar(CH$_2$)$_2$Ar (including hydroaromatic) and Ar-CH$_2$-Ar (Ar - aromatic ring) units respectively. Examination of the concentrations of these major aliphatic products suggests that the medium lithotypes contain the highest concentrations of aryl methyl and Ar(CH$_2$)$_2$Ar structures. Diaryl methane units are proposed to be most significant in the pale lithotype possibly reflecting a resin input.

Examination of the Neutrals, Acids 1 and Acids 2 contributions to the oxidation products (Table 1) reveals that their concentration varies consistently in that Acids 2 > Acids 1 > Neutrals. The fractions exhibit a lithotype dependence in their concentrations with the lighter lithotypes reporting higher values, particularly in the less polar Neutrals and Acids 1 fractions.

The light and medium dark lithotypes are widely separated in their oxidation product yields (Table 1). Chromatographic separation and analysis of their Neutrals, Acids 1 and Acids 2 fractions generally resulted in the widest variation in structural type and yield for the lithotype series. The results of these analyses are depicted in Tables 2 and 3 along with Figure 6A, B and C.

Neutrals Fraction: Both chromatograms (Figure 6A) are dominated by straight chain hydroxylated species (Table 2). The chromatograms are qualitatively similar although quantitatively (Table 3) the litho-

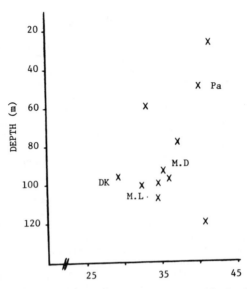

HYDROGEN WT% YIELD c.f. INITIAL COAL H. CONTENT

Figure 5   Plot of pertrifluoroacetic acid oxidation aliphatic
            proton yield data derived from P.M.R. and elemental
            analysis.

Table I.   Product Composition of Brown Coal Lithotypes Pertrifluoro-
           acetic Acid Oxidation Mixtures

Lithotype	Residue	Acids					Neutrals	Total
		Acetic	Succinic	Malonic	I	II		
Pale	0.75	3.80	2.07	3.38	2.62	29.47	0.42	42.51
Light	2.53	4.80	2.36	1.82	2.68	26.85	0.43	41.47
M. Light	0.54	4.89	3.19	1.47	1.51	28.55	0.32	40.47
M. Dark	0.52	5.08	3.19	1.47	1.29	24.27	0.07	35.89
Dark	1.34	3.96	1.90	1.67	1.28	21.90	0.13	32.18

- Weight % of initial coal lithotype appearing as oxidation product.

Table II. Pertrifluoroacetic Acid Oxidation Product Distribution

Fraction	Compound Class	LIGHT LITHOTYPE			MEDIUM DARK LITHOTYPE		
		Distribution*	Maxima	Even/Odd Predominance	Distribution*	Maxima	Even/Odd Predominance
Neutrals	n-alcohols	C_9-C_{37}	C_{15},C_{26}	$C_{20}=<; C_{20}++>$	C_9-C_{37}	C_{17},C_{30}	$C_{20}=<; C_{20}++>$
"	α,ω-diols	C_6-C_{32}	C_{12},C_{19}	$C_{16}+<; C_{18}->$	C_6-C_{32}	C_{12},C_{21}	$C_{16}+<; C_{18}->$
"	(ω-1→ 6)-diols	C_7-C_{16}	C_{11}	=	C_7-C_{16}	C_{12}	=
Acids-1	n-monocarboxylic acids	C_8-C_{33}	C_{16},C_{28}	++	C_8*-C_{34}	C_{12},C_{28}	++
"	ω-OH, n-mono-carboxylic acids (ω-1→ -6)-OH,	C_7-C_{27}	C_9	$C_{20}=<; C_{20}++>$	C_8-C_{24}	C_{10}	=
	n-monocarboxylic acids	C_9-C_{20}	C_{12}	+	C_8-C_{20}	C_{11}	+
	α,ω-n-dicarboxylic acids	C_9-C_{27}	C_{12}	=	C_8-C_{27}	C_{13}	=
	γ-lactones 4-hydroxy-n-carboxylic acids	C_6-C_{10}	C_8	=	C_6-C_{10}	C_9	=
Acids-2	α,ω-n-dicarboxylic acids	C_3-C_9	C_5*	=	C_3-C_9	C_5*	=
	Tricarboxylic acids	C_2-C_5	C_3	=	C_2-C_5	C_3	=

- odd/even predominance   + even/odd predominance

= no distinct predominance   ++ strong even/odd predominance

* Isolation methods employed preclude the accurate analysis of members with lower carbon number.

120   THE CHEMISTRY OF LOW-RANK COALS

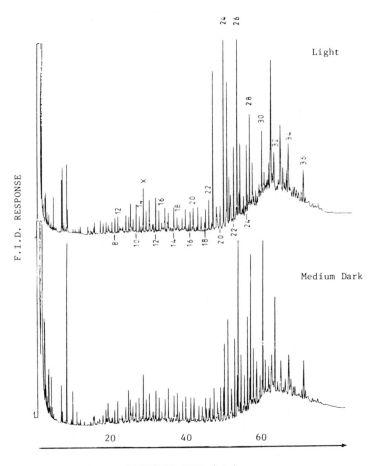

RETENTION TIME (min)

Figure 6A   Chromatographic separation, of the Neutrals fractions
             from the Light and Medium Dark Lithotypes. Numbers
             directly above eluted peaks identify the chain length
             of these even n-alcohols. Numbers below the chromato-
             gram refer to the carbon chain length of the correspond-
             ing terminally hydroxylated diols. (X) is identified
             as 1,9-dihydroxy-dodecane.

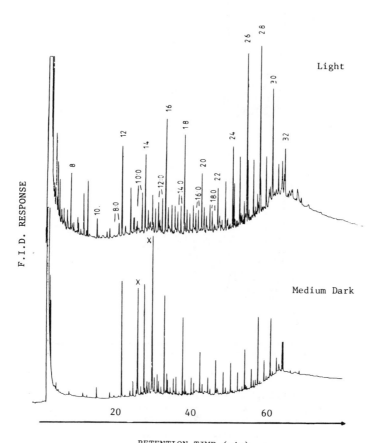

RETENTION TIME (min)

Figure 6B   Chromatographic separation of the Acids 1 fraction from
the Light and Medium Dark Lithotype. Numbers directly
above eluted peaks identify the carbon chain length of
these even monocarboxylic acids. Numbers ending in :0
identify the carbon number of the (α,ω) dicarboxylic
acid and later eluting ω-hydroxy, monocarboxylic acid.
Peaks marked with an X are lactones.

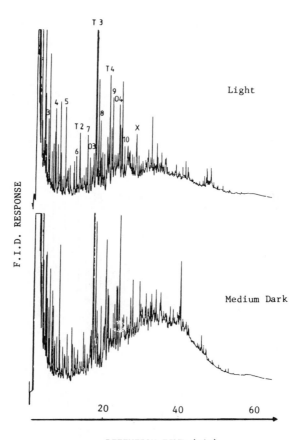

<u>Figure 6C</u>   Chromatographic separation of the polar Acids 2 fraction.
            Numbered peaks refer to the carbon chain length of (no
            prefix) dicarboxylic acids, (T prefix) tricarboxylic
            acids and (0 prefix) oxirane acids.   X identifies 1,2-
            epoxy-1,1,2,3-propanetetracarboxylic acid.

Table III.   Weight %* Composition of Product Classes

	Alcohols	Mono Acids	Di Acids	Tri Acids
Pale	0.14	1.78	0.40	0.45
Light	0.31	1.68	0.42	0.73
M. Light	0.21	1.27	0.49	1.07
M. Dark	0.02	0.69	0.34	0.54
Dark	0.10	0.84	0.29	0.42

*Relative to parent coal lithotype

types are quite distinct. Straight chain n-alcohols dominate the chromatograms, terminally hydroxylated (α, ω) diols present the second most important series followed by diols containing random positioning of the second -OH group between ω-1 and ω-6.

The neutral material is thought to derive primarily from molecules trapped within the macromolecular coal matrix (13). This theory is supported by the structural independence exhibited by the neutral chromatograms to variations in coal lithotype. The presence of multiple hydroxyl substitutional isomers is thought to reflect oxidation by the reagent mixture.

Acids 1 Fraction: Both chromatograms (Figure 6B) are dominated by series of straight chain carboxylic acids. The light lithotype distribution being characterized by C_{24}-C_{30} monocarboxylic acids while, in contrast, the medium dark lithotype produces significant concentrations of two lactones which are absent in the light fraction. The precise structure of these lactones is unknown; however, their mass fragmentograms are consistent with a 5-OH (σ-lactone) and 4-OH (γ-lactone) long chain methoxy diacids. The Acids 1 fraction of the light lithotype contains significantly more benzoic and phthalic acids than the medium dark despite the lower overall aromaticity reported for the parent lithotype (Figure 2). These structural differences are speculated to reflect a high resin contribution to the light lithotype structure in comparison to a high lignin input for the medium dark. The Acids 1 fraction is considered to originate primarily from substituents on the coal matrix. This proposed origin is consistent with the structural composition of the Acids 1 fraction being affected by lithotypic variation.

Acids 2 Fraction: The identified products within this fraction comprise short chain di- and tricarboxylic acids with similar distributions present for all lithotypes. The general complexity of the chromatograms (Figure 6C) results from the multitude of different isomers comprising hydroxylated versions of these polyacids. Aromatic tri- and tetra-carboxylic acids are present in minor concentrations within these fractions. A distinguishing feature of the medium dark Acids 2 fraction is the large contribution of unresolved components producing a distinct hump in the chromatogram (Figure 6C). The comparatively low elution temperature (< 250^{0}C) and dominance of aromatic and trimethyl silyl fragment ions in single ion GC/MS plots suggests the "hump" contains highly polar low molecular weight aromatic moieties. These molecules would be protected from further oxidation by the reagent due to the nature of these substituents (13). The presence of this unresolved material may relate to a high lignin derived structural contribution to the medium dark lithotype. The similarity in the distribution of identifiable molecules between the different lithotype Acids 2 fractions suggests a common higher plant origin for their skeletal structures.

Characterization of Coal Fractions

The variation in the weight percentage of total solvent extractable (TSE) material (extraction with chloroform/methanol/toluene) and solvent extractable humic acid (SEHA) for the coal samples is illustrated in Figure 7. There is a substantial reduction in the weight percentage of these fractions in the progression from paler to darker lithotypes. It can also be seen that the variation with lithotype is significantly more extreme than the variation within the suite of *light* lithotypes. Within the latter suite there is a notably large difference in the weight percentages between the two adjacent samples from 59 m and 78 m depth in the core. The jump across this interval can be related to the Morwell 1A/Morwell 1B seam boundary which occurs at the 73 m depth. Taking into account this abberation the data are consistent with an increase in extractable material as a function of depth (for samples within a particular lithotype classification and seam). The increase across the interval from 78 to 119 m depth is quite marked by comparison with the increase in extractable material which accompanies increasing rank within the subbituminous rank regime (7). There have been, however, no previous systematic studies giving any indication of what to expect within this lower rank interval. A more intense study of rank dependent phenomena within this interval could significantly enhance our understanding of coal forming processes.

The distributions of extractable n-alkanes from the different lithotypes are illustrated in Figure 8. All the distributions exhibit a marked predominance of odd carbon chain length homologues and in this respect all are similar to the distribution of n-alkanes which occur in living higher plant waxes. In other respects, however, the distributions differ markedly. For example, the *dark* lithotype exhibits a primary maximum at C_{29} and a secondary one at C_{25}. The *medium* dark and *pale* lithotypes exhibit only primary maxima at C_{31} and C_{29}, respectively. By contrast, all the *light* samples exhibit primary maxima at C_{29} and secondary maxima at C_{37}. These C_{37} maxima appear to be a distinctive feature of the *light* samples and have almost certainly resulted from the input of some fairly specific taxon which was viable only in association with the paleoenvironment of deposition for this lithotype. The distribution can be further characterized by Carbon Preferences Indices (CPIs), a parameter which is effectively the ratio of the abundance of odd relative to even carbon n-alkane chain lengths. It can be seen (Figure 9) that there is considerable spread in the CPI values for the *light* samples and that only the CPI value for the *dark* lithotype lies significantly outside this margin. There is no apparent change in the distribution of n-alkanes as a function of depth and it would appear that the distributions remain substantially representative of those contributed by the higher plant progenitors of the coal.

A consideration of the extractable n-fatty acid distributions gives a complimentary characterization of the sample suite. The distributions are again a reflection of those observed in the higher plant precursors to the coal. The distributional differences between lithotypes and the similarities within a single lithotype classification are less obvious than in the case of the n-alkanes, but a statistical approach involving the determination of covariance between sample pairs was able to effectively differentiate the divergent nature of the distributional variations as a function of depth and lithotype (5). CPI values for the n-fatty acid distributions are plotted in Figure 10 (the CPI, in this case specifically for fatty acids, is the ratio of even relative to odd chain lengths).

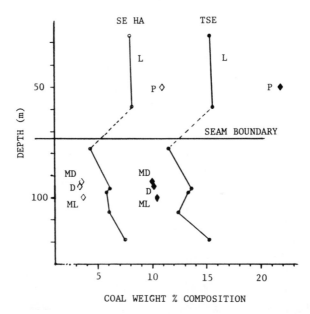

Figure 7   Weight % (closed symbols) and SEHA (open symbols) versus
depth.   The horizontal line at 73 m represents the
Morwell 1A/Morwell 1B seam boundary.

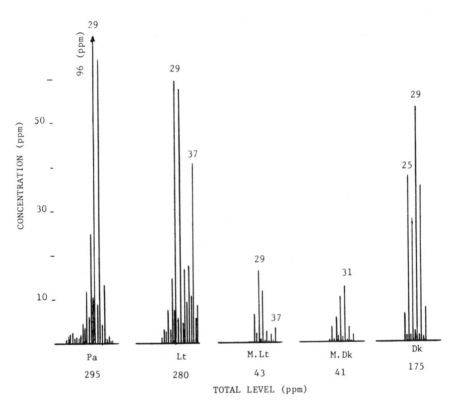

<u>Figure 8</u>   Distribution of n–alkanes as a function of lithotype.

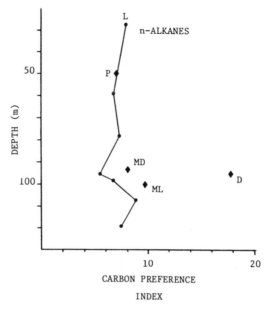

Figure 9   Carbon Preference Indices of n-alkanes versus depth.

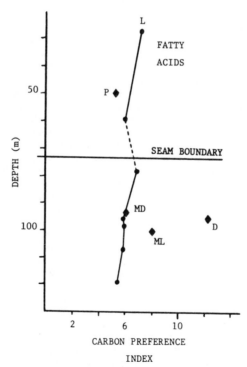

Figure 10   Carbon Preference Indices of n-fatty acids versus depth.

Again there is a considerable difference in the values for the
various lithotypes, but the spread of values within the *light* sample
suite is relatively small. Within this spread however, it is
possible to discern a general reduction in the CPI values with depth.
There is again a discontinuity across the seam boundary. There is a
corresponding general reduction in the level of extractable n-fatty
acids over this depth interval and the consistent variation in these
two parameters is believed to indicate that progressive diagenetic
removal and alteration of the fatty acids is occurring with depth
within this coal sequence. The discontinuity across the seam
boundary is an indication of the complex combination of paleobotanical, geochemical and microbiological factors which control the
diagenetic progression.

Many coals contain extended chain (> 31 carbon atoms) hopanes
and related compounds which appear to be diagenetically derived from
the lipids of certain procaryotic organisms. Biologically produced
hopanoids exclusively possess $17\beta H$, $21\beta H$ stereochemistry. The

$17\beta H,21\beta H$-homohopane

thermodynamically more stable $17\alpha H,21\beta H$ stereochemistry has usually
been explained to form in the geosphere via diagenetic acid catalyzed
and temperature dependant isomerization from the biological diastereomers. Hence the diastereomeric ratio, $17\alpha H,21\beta H/17\beta H,21\beta H$ has been
observed to rise in simulated maturation studies and with increasing
depth of burial in sedimentary sequences. The present data exhibit
a reversal of this trend; that is, there is a reduction in this ratio
with depth for the series of *light* samples (Figure 11). Although
there must be some mechanism to account for the formation of the
abiological $17\alpha H,21\beta H$-diastereomers in the first place, the data
suggest that $17\beta H,21\beta H$-diastereomers are being progressively produced
with depth in the coal seam. On the basis of present understanding
this could only occur by microbiologically mediated processes. The
temperatures reached within the coal seam are obviously insufficient
to force the isomerization towards its thermodynamic equilibrium.
There is also a significant difference in the diastereomeric ratios
determined by lithotype. It is probable that this marked difference
is substantially determined during the early stages of coal deposition where it may be significantly influenced by the pH of the
surrounding swamp water or the coal surface acidity. The significant
difference in the diastereomeric ratios (Figure 10) as a function of
lithotype is consistent with the preceeding explanation; that is, the
dark lithotypes can be generally associated with higher acidities
(and hence lower pH) which would favour the formation of the
abiological isomer, as observed.

The polycyclic aromatic hydrocarbons (PAHs) in brown coal consist
predominantly of pentacyclic angularly condensed hydroaromatic
moieties. These compounds appear to be derived from C_{30} triterpenoid

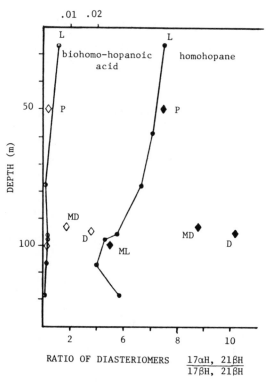

Figure 11   Ratio of diastereomers, 17αH,21βH/17βH,21βH, for homohopane (closed symbols, bottom axis) and bishomo-hopanoic acid (open symbols, top axis).

precursors which occur as natural products in angiosperm waxes.   It
can be seen (Figure 12) that there is a marked variation in the
concentration of PAHs as a function of both lithotype and depth.   The
variation as a function of lithotype is consistent with our views on
the generalized paleoenvironments of deposition, i.e., the lighter
lithotypes are deposited in relatively anaerobic environments more
conducive to the preservation of the originally deposited organic
material, while the darker lithotypes are deposited in relatively
aerobic environments in which much of the less resistant organic
material is removed by oxidation or substantially altered microbiologically.   The increasing concentration in PAH concentration with
depth for the series of *light* samples suggests that the formation of
PAHs is also occurring within the coal seam.   It seems most likely
that this transformation, too, is microbiologically mediated.

Conclusions

The organic chemical nature of Victorian brown coal has been shown to
vary as a function of both depth and lithotype.   In the latter case
the structural differences occur predominantly as a result of the
differing nature of the depositional paleoenvironments and are
readily distinguishable by both gross and molecular level parameters.
The differences with depth are considerably more subtle and were
observed only by specific molecular techniques; e.g., the level of
PAHs, the diastereomeric ratio of hopanoids and the CPIs of the
n-fatty acid distributions.   Since the observed differences at the
molecular level involve only a small proportion of the total coal
substrate (parts per thousand level) it is hardly surprising that
they are not manifested at the gross structural level.   Although mild

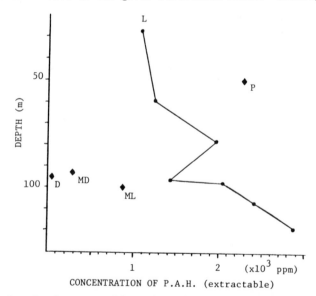

Figure 12   Level of extractable polycyclic aromatic hydrocarbons
with depth.

increases in temperature certainly occur across the sampled coal interval, the specific nature of the chemical changes observed with depth suggests that microbiological processes rather than geochemical (temperature dependent) processes are chiefly responsible. It may be that the warmer conditions deeper in the seam favour the viability of anaerobic microorganisms which can modify the coal substrate in the required manner, or more probably that the closer proximity of the deeper samples to underground aquifers (and potential sources of oxygen and inorganic nutrients) favour the viability of aerobic microorganisms.

Acknowledgments

The authors acknowledge the Victorian Brown Coal Council and the State Electricity Commission of Victoria for the provision of coal samples. Mr S. Bombarci is thanked for technical assistance. J. Espitalié and G. Maciel are thanked for providing access to Rock-Eval and solid state NMR facilities. G. Eglinton and J. K. Volkman are thanked for access to GC/MS facilities.

Literature Cited

1. Allardice, D. J.; George, A. M.; Häusser, D.; Neubert, K. H. and Smith, G. C.; State Electricity Commission of Victoria, Research and Development Department, Report No. 342, 1977.
2. Blackburn, D. T.; presented at "Cainozoic Evolution of Continental South-Eastern Australia" Conference, Canberra, Nov. 1980.
3. George, A. M.; State Electricity Commission of Victoria, Petrographic Report No. 17, 1975.
4. Deno, N. C.; Greigger, B. A.; Stroud, S. G. Fuel 1978, 57, 455-459.
5. Johns, R. B.; Verheyen, T. V. and Chaffee, A. L. "Proceedings of the International Conference on Coal Science, Düsseldorf, 1981"; Verlag Glückauf, Essen, 1981, pp. 863-868.
6. Chaffee, A. L.; Perry, G. J.; Johns, R. B. and George, A. M. In "Coal Structure"; Gorbaty, M. L.; Ouchi, K., Eds.; ADVANCES IN CHEMISTRY SERIES No. 192, Washington D.C., 1981; Ch. 8, pp. 113-131.
7. Verheyen, T. V. and Johns, R. B. Geochim. Cosmochim. Acta 1981, 45, 1899-1908.
8. Chaffee, A. L. and Johns, R. B. Geochim. Cosmochim. Acta 1983, 47, 2141-2155.
9. Robin, P. L. and Rouxhet, P. G. Geochim. Cosmochim. Acta 1978, 42, 1341-1349.
10. Miknis, F. P.; Sullivan, M.; Bartuska, V. T. and Maciel, G. E. Org. Geochem. 1981, 3, 19.
11. Espitalié, J. Rev. Inst. Fr. Petrole 1977, 32, 32-42.
12. Battaerd, H. A. J. and Evans, D. G. Fuel 1979, 58, 105-108.
13. Verheyen, T. V. and Johns, R. B. Anal. Chem. 1983, 54, 1564-1568.
14. Radke, M.; Schaefer, R. G.; Leythauser, D. and Teichmuller, M. Geochim. Cosmochim. Acta 1980, 44, 1787-1800.

RECEIVED April 1, 1984

Some Structural Features of a Wilcox Lignite

NARAYANI MALLYA and RALPH A. ZINGARO

Department of Chemistry and the Coal and Lignite Research Laboratory, Texas A&M University, College Station, TX 77843

A hypothetical model of a Wilcox (Texas) lignite is proposed based on data obtained from the following experimental methods: ^{13}C CP/MAS (cross-polarization/ magic angle spinning) NMR, ^1H NMR of the Me_2SO extract of the lignite acidic group determination, pertri-fluoroacetic acid oxidation and positive identification of some liquefaction products. The aromaticity (f_a) value is found to be 0.53. Other features characteristic of the structure of this lignite are (i) the predominance of longer chain aliphatic substituents on the aromatic rings; (ii) the apparent absence of methylene bridges; (iii) fused aromatic ring systems rarely exceeding two rings; (iv) a ratio of carboxylate to carboxylic acid groups > 1; and (v) the presence of phenolates along with phenolic groups and a very low concentration of methoxy groups.

The state of Texas is one of the largest producers and consumers of lignite in the nation. It has reserves estimated at 23.4 billion tons of near surface deposits occurring in three formations, viz., Wilcox, Jackson and Yegua (1). Geologically, 71% of the total tonnage is estimated to reside with the Wilcox group. Based on the Btu content, the Wilcox group accounts for 77% of the total (1). At the present time four lignite fired power generating plants are operating in the state and several more are under construction. In addition to the near-surface reserves, extensive deposits of deep-basin lignite also exist, but they have been little investigated. An estimated 35 billion tons of deep-basin lignites exist (1) and may be partially recoverable using in-situ gasification, deep surface mining or underground mining techniques. Of the total tonnage, 69% is in the Wilcox formation. In order to utilize this resource more efficiently, knowledge about the chemical constitution of these lignites is useful. Very little information is currently available about the structure of Texas lignites.

This study was undertaken for the purpose of obtaining some ini-

0097–6156/84/0264–0133$06.00/0
© 1984 American Chemical Society

tial information concerning the structural features of a lignite
sample from the Wilcox (Texas) group. Experimental methods utilized
included ^{13}C CP/MAS NMR, ^{1}H NMR, acidic group determination, pertrifluoroacetic acid oxidation and identification of some liquefaction
products obtained from this lignite.

Experimental

The lignite sample used in this study is from Rockdale, located in
Freestone County in Central Texas. This lignite belongs to the
Wilcox formation and is of deltaic origin. This sample was donated
by the Dow Chemical Company, U.S.A. after storage. The sample was
ground to -100 mesh under nitrogen and stored in containers under an
atmosphere of nitrogen. Proximate analyses were done using a Fisher
Coal Analyzer, Model 490. Elemental analyses of the neat lignite and
demineralized lignite sample were performed by the Huffmann Laboratories, Inc., Wheatridge, Colorado. Demineralization was accomplished by treatment with HF/HCl as described by Bishop and Ward (2)
and complies with the International Standards set forth for the
direct determination of oxygen in coals of high ash content (3).

^{13}C and ^{1}H NMR Studies. ^{13}C CP/MAS NMR spectra were measured by Dr.
B. Hawkins and Dr. J. Frye of the Colorado State University, NMR
Center. A Jeol FX 60Q spectrometer was used. The experimental
conditions used were the following: ^{13}C frequency – 15.04 MHz;
decoupling frequency – 59.75 MHz; number of scans – 40432; spectral
width – 8000 Hz; sample state – solid; reference – external hexa-
methylbenzene.
     The proton NMR spectrum of the Me_2SO extract of the lignite was
obtained using a XL-200 Fourier transform proton NMR spectrometer.
Details of lignite extraction with Me_2SO are described elsewhere
(4,5). Instrumental conditions for the measurement of the spectrum
were as follows: frequency – 200 MHz proton; sample state – liquid
solution; solvent – Me_2SO-d_6 locked; sample concentration – ca. 200
mg/10 ml; probe temperature – 25°C; sweep width – 2600.1 Hz; acquisition time – 1 second; internal reference – tetramethyl silane; number
of transients – 1500; pulse width – 5.0 µseconds.

Acidic Group Determination. The concentration of carboxyl groups and
total acidity of the lignite were obtained using the method described
by Schafer (6). Concentrations of carboxyl groups and the total concentration of acidic groups were also determined following demineralization. Demineralization was carried out according to the method of
Teo, et al. (7). This procedure involves treatment of the sample
with 5N HCl to ensure that all acid groups present are converted to
the free acid form rather than in the form of metal-bound (carboxylate and phenolate) cations. The concentration of carboxyl groups
and total acidic group concentration of the demineralized sample was
obtained by the two different methods described by Schafer (6).
Carboxylic group determination consisted of ion exchange with calcium
acetate and back titration of the liberated acetic acid or back-
exchange of the coal treated with calcium acetate with HCl and titration of the excess HCl. Total acidity was determined by barium
hydroxide exchange and determination of the residual barium hydroxide

or by back-exchange of the barium salt with acid and determination of the acid consumed. A drop of surfacant solution (Surfynol) was used in all the titrations to keep the coal particles in suspension and prevent their floatation to the surface of the titrating medium. A Ross pH electrode was used to measure the pH of the solutions.

Pertrifluoroacetic Acid Oxidation. Pertrifluoroacetic acid oxidation was introduced by Deno and coworkers (8) for the study of aliphatic linkages and substituents in coal structure. Shadle and co-workers (9) modified this procedure. In this study the modified procedure was used. One gram of dried -100 mesh lignite was initially refluxed with 40 ml of chloroform in a round-bottomed flask. The refluxed mixture was allowed to cool. Pertrifluoroacetic acid was prepared by mixing trifluoroacetic acid, 30% H_2O_2 and concentrated H_2SO_4 in the ration of 8/10/5:v/v/v (ml). Forty ml of the acid mixture was added to the chloroform-lignite mixture and refluxed for 5 h. The reaction mixture was then cooled in ice. About 20 mg of 5% Pt/asbestos was added and the mixture was stirred overnight to decompose any excess of H_2O_2. The mixture was then filtered through a pad of glass wool to remove the black residue, ca. 10% of sample. The pad was rinsed thoroughly with $CHCl_3$. The $CHCl_3$ layer was separated and dried over anhydrous $MgSO_4$. It was filtered and then evaporated at reduced pressure to remove the solvent. A small portion of the residue was taken up in $CDCl_3$ for proton NMR measurement. The remainder of the reaction product was esterified by the addition of 2 ml of 14% BF_3 in methanol for every 0.1 g of the starting coal oxidation product and refluxed for 2 h. This converted the carboxylic acids formed in the reaction to their methyl esters. The esters were recovered by extraction with methylene chloride. The methylene chloride extract was washed with 10% Na_2CO_3 and the solvent was removed. The esters were dissolved in ether and analyzed by GC-MS using an SE-30 capillary column having a length of 28 m and temperature program of 50-250°C and a heating rate of 4°/min.

Liquefaction Study. The lignite was liquefied using the method described by Phillip and Anthony (10). The liquefied product was extracted with Me_2SO as described by Mallya (4) and the extract was characterized by GPC (gel permeation chromatography) separation and GC-MS as described by Zingaro and co-workers (11).

Results and Discussions

Table 1 shows the proximate analyses and elemental analyses of the lignite. The table also shows elemental analyses of the demineralized lignite.

Figure 1 shows the ^{13}C CP/MAS NMR spectrum of the neat lignite. The aromaticity (f_a) value calculated from the spectrum is 0.53. Since the Me_2SO extract obtained under ambient conditions has been shown to possess chemical and physico-chemical properties very similar to those of the bulk lignite (4,5), an approximate estimate of the proton distribution in the lignite is obtained from the proton NMR spectrum of the Me_2SO extract of the lignite. Table 2 shows the proton population of the Me_2SO extract. The proton NMR spectrum of the Me_2SO extract has been divided into six regions as described by

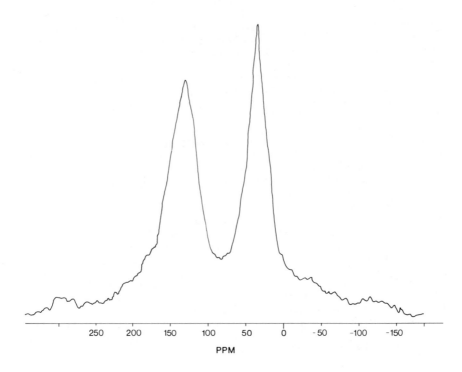

Figure 1. ^{13}C CP/MAS NMR Spectrum of neat Wilcox (Texas) Lignite.

Table 1.   Proximate Analyses and Elemental Analyses of Wilcox Lignite

Proximate Analyses	%	Elemental Analyses	Neat % daf	Demineralized % daf
Moisture	29.26	Carbon	73.10	70.65
Volatile matter	46.66 mfb	Hydrogen	5.18	5.02
Ash	12.55 mfb	Oxygen	22.58	19.62
Fixed carbon	48.78 mfb	Nitrogen	1.49	1.48
		Sulfur	1.54	1.97
		Halogen	0.97	0.78

Brown and Ladner (12) and Ladner et al. (13). The six regions are described in Table 2. As can be seen from the table the total popu-

Table 2.   Proton Population (Area %) in Me_2SO Extract of Wilcox Lignite

Proton Type	Symbol used	δ(ppm)	Area %
Terminal methyl groups	$H_{(\gamma)}$	0.5–1.0	13.7
Methylenes β to or further than β-position from polar groups or aromatic rings	$H_{(\beta)}$	1.0–1.5	20.8
Napthenic methylene groups	$H_{(N)}$	1.5–2.0	12.0
Methylenes α to carboxylic or carbonyl groups or aromatic rings	$H_{(\alpha 1)}$	2.0–3.3	24.0
Methylenes α to ether groups	$H_{(\alpha 2)}$	3.3–4.5	14.2
Aromatic ring + phenolic protons	$H_{(Ar)}$	4.5–9.0	15.3

lation of aliphatic protons, i.e., $H_{(\alpha)}$ + $H_{(\beta)}$ is higher than the $H_{(Ar)}$ population. There is also a high population of $H_{(\alpha 1)}$ and $H_{(\alpha 2)}$ suggesting a high concentration of oxygen functionalities in the lignite.

The proton NMR data, together with the ^{13}C CP/MAS NMR spectrum and the elemental analyses have been used to calculate two important structural parameters, σ and H_{aru}/C_{ar}, based on the equations of Brown and Ladner (12). The parameter, σ, is defined as the degree of substitution of the aromatic systems, i.e., the fraction of the available ring carbon atoms which bear substituents and H_{aru}/C_{ar} is defined as the ratio of hydrogen to carbon in an hypothetically unsubstituted aromatic material. The value of σ is found to be 0.8. This value is very close to that expected for a napthalene system [the value for benzenoid structure is 1; for napthalene 0.8 and for three-ring systems 0.72 (14)]. The lignite appears to contain very low concentrations of fused aromatic ring structures greater than two.

From the relationship of van Krevelen and Schuyer (15) and the equations of Williams and Chamberlain (16) structural parameters such as ring condensation index 2 [(R-1)/C]; average number of total rings per molecule (R); average number of aromatic rings per molecule (R_a); average number of napthenic rings per molecule (R_n); number of carbon atoms in the aromatic rings (C_a); number of carbon atoms in the saturated groups (C_s) and number of carbon atoms in the substitutable aromatic ring carbons (C_l) are calculated for the lignite. The formulae used in the calculations are shown in Table 3 and the structural parameters are shown in the same table. Values calculated by the use of these equations are useful and highly suggestive, but they cannot be considered to be quantitative.

Concentrations of carboxyl groups and total acidity for the lignite before and after demineralization are shown in Table 4. Concentrations of phenolic groups, carboxylate groups and phenolate groups are calculated from the titration data and are shown in Table 4. As is noted from Table 4 the lignite has a higher concentration of carboxylate as compared with carboxylic acid groups. Carboxylate concentrations are of interest since the higher reactivity of low-rank coals and their chars is attributed to the catalytic effect of the inorganic constituents present, especially the ion-exchangeable cations (17). Coal pyrolysis gives rise to volatile matter and char. Char is of considerable practical importance because the rate of char combustion is considered to represent the rate-determining step for the overall combustion process. Char is also important in gasification processes. Chars from low-rank coals, such as lignites, have been found to be more reactive (18) than chars produced from high rank bituminous coals. According to Morgan and co-workers (19) American lignites are rich in alkali and alkaline-earth metal cations. Walker (20) and McKee (21) have shown that alkali and alkaline-earth metals are good catalysts for the carbon-oxygen reaction. Hence, the high concentrations of carboxylates in Texas lignites represent factors which could have considerable practical importance. The presence of phenolate groups in the lignite is also indicated in Table 4. The effects of phenolates in coal conversion processes have not been studied in any detail, but they may have effects similar to those of the carboxylates. The proton population in the pertrifluoroacetic acid oxidation product of the lignite is shown in Table 5. The high population of $H_{(\beta)}$ suggests the presence of longer chain aliphatic substituents and bridges in the lignite. According to Deno and co-workers (8) acetic acid is formed from molecules such as toluene, dimethylbenzenes, 2-methylnapthalene and, in smaller quantities, from ethylbenzene, propylbenzene, isopropylbenzene, propylphenols, benzyl acetate and 1-phenylethanol. The very low population of acetic acid protons in the oxidation products obtained from the lignite which is the subject of this study suggests the presence of low concentrations of short-chain alkyl substituents. The low population of adipic or suberic or sebacic acid protons suggests small quantities of C_4, C_6 or C_8 alkyl bridges between aromatic

Table 3. Equations Used in the Calculations of Various Structural Parameters of Wilcox Lignite and Resultant Values

Structural Parameters	Equations	Value
f_a	$\dfrac{C_{total} - C_{aliphatic}}{C_{total}}$	0.53
σ	$\dfrac{\dfrac{H_\alpha^* + O/H}{x}}{\dfrac{H_\alpha^* + O/H + H_{ar}^*}{x}}$	0.71
H_{aru}/C_{ar}	$\dfrac{H_\alpha^* + H_{ar}^* + O/H}{C/H - H_\alpha^*/x - H_o^*/y}$	0.84
R	$[C-H/2 - Cf_a/2] + 1$	2.80
C_a	Cf_a	3.11
C_s	$C-C_a$	2.76
R_a	$\dfrac{(C_a-2)}{4}$	0.28
R_n	$(R-R_a)$	2.52
C_l	$(C_a/2) + 3$	4.56
$2[R-1/C]$	$2-f_a-H/C$	0.62

ring structures. The low population of butyric acid protons suggests the presence of small quantities of phenyl–C_3 type systems.

Additional information concerning the length of alkyl bridges and substituents is obtained from GC/MS data. The long-chained acid esters identified show aliphatic substituents from C_7–C_{17} and aliphatic bridges of C_{10}, C_{11}, and C_{15} lengths. The long-chained acid esters identified are shown in Table 6. Many compounds remain unidentified in the GC–MS since the spectra are too complicated to be interpreted. Also, many of the product components may have been lost in the work-up procedure due to lack of solubility in the solvent

Table 4.   Concentration of Acidic Groups and Metal-Bound Acid Groups
in Wilcox Lignite

Functional Group	meq/g daf.
Carboxyl groups in neat lignite[a]	0.44
Total acidity of neat lignite[a]	3.10
Phenolic groups in neat lignite = total acidity - carboxyl groups	2.66
Carboxyl groups in demineralized lignite[b]	2.12
Total acidity of demineralized lignite[b]	6.60
Phenolic groups in demineralized lignite = total acidity - carboxyl groups	4.48
Concentration of carboxylates = carboxyl$_{demin.}$ - carboxyl$_{neat}$	1.68
Concentration of phenolates = phenol$_{demin.}$ - phenol$_{neat}$	1.82
Total cations present = total acidity$_{demin.}$ - total acidity$_{neat}$	3.50

[a] Obtained by determination of base consumed in exchange
[b] Average of values obtained by a and those obtained by determination of acid consumed in back-exchange of exchanged coal

Table 5.   Proton Population (Area %) in the Pertrifluoroacetic Acid
Oxidation of Products of Texas Lignite

Proton Type	δ(ppm)	Area %
$H_{(\gamma)}$	0.5-1.0	15.79
$H_{(\beta)}$	1.0-1.5	65.79
$H_{(N)}$	1.5-2.0	7.89
acetic acid	2.06	2.63
adipic, suberic or sebacic acid	2.21	2.63
butyric acid	2.29	5.26

Table 6. Methyl Esters of the Long Chained Acids in the Oxidation
Products of Texas Lignite[a]

(1)	octanoic acid ester
(2)	decanoic acid ester
(3)	decanoic acid ester with ethyl substitution at various C-atoms
(4)	undecanoic acid ester
(5)	undecanoic acid ester with methyl substitution at various C-atoms
(6)	dodecanoic acid ester
(7)	dodecanoic acid ester with methyl substitution at various C-atoms
(8)	tridecanoic acid ester
(9)	tetradecanoic acid ester
(10)	pentadecanoic acid ester
(11)	hexadecanoic acid ester
(12)	heptadecanoic acid ester
(13)	heptadecanoic acid ester with methyl substitution at various C-atoms
(14)	octadecanoic acid ester
(15)	dodecadioic acid diester
(16)	tridecadioic acid diester
(17)	heptadecadioic acid diester

[a] Tentatively identified by matching each individual spectrum with
the authentic spectrum

used and some compounds are not detected by GC/MS due to lack of vol-
atility of components.   In addition to the long-chain acids, some
terpenes and isoprenoid structures, as indicated by the fragmentation
patterns, are observed, but not unequivocally identified.

From the products positively identified, it is obvious that
Texas lignites possess longer aliphatic chains attached to and
joining the aromatic rings. A geological and geochemical interpretation of the presence of long aliphatic chains in this lignite indicates that it has both a terrestrial and an aquatic plant ancestry.
Terrestrial plant debris contains a higher content of aromatic structures whereas aquatic organic debris (algae, plankton, etc.) furnish
more aliphatic structures. The particular lignite which is the subject of this research has been classified as deltaic by geologists
and consequently, should have considerable contribution from the
marine organisms. This would explain the presence of long aliphatic
chains. However, contributions to the structure which have their
origin in cellulose cannot be ruled out since structures of cellu-
losic origin have been found in many brown coals.

Further evidence for the presence of alkyl substituents as long
as C_{12} is obtained from the liquefaction study. The products
obtained by liquefication of the lignite which are soluble in Me_2SO
show the presence of a series of alkyl phenols (Table 7). The Me_2SO
solubles of the liquefied lignite also show some C_3-alkyl binaptha-

Table 7. Alkyl Phenols Tentatively Identified in Liquefied
Texas Lignite

Compounds	Retention Time, min.
(1) phenol	2.4
(2) o-cresol	3.8
(3) p-cresol	4.4
(4) m-cresol	4.6
(5) C_2-alkyl phenol	6.0
(6) C_2-alkyl phenol	6.2
(7) C_2-alkyl phenol	6.9
(8) xylenol	7.2
(9) xylenol	7.7
(10) xylenol	7.8
(11) xylenol	8.0
(12) C_3-alkyl phenol	8.4
(13) C_2-alkyl cresol	8.6
(14) C_2-alkyl cresol	9.0
(15) C_3-alkyl cresol	9.6
(16) C_3-alkyl cresol	9.7
(17) C_3-alkyl cresol	9.9
(18) C_3-alkyl cresol	11.2
(19) C_4-alkyl cresol	12.4
(20) C_4-alkyl cresol	12.7
(21) C_4-alkyl cresol	13.6
(22) C_9-alkyl phenol	15.4
(23) C_9-alkyl phenol	15.5
(24) C_9-alkyl phenol	16.2
(25) C_9-alkyl phenol	16.3
(26) C_9-alkyl phenol	16.8
(27) C_9-alkyl phenol	17.0
(28) C_{12}-alkyl phenol	17.2
(29) C_9-alkyl tri-phenol	18.1

lenes. There are very few polynuclear aromatics that have been thus
far detected in the liquid. This is consistent with the H_{aru}/C_{ar}
value previously mentioned which suggests a very minor, if any presence of larger ring systems.

Conclusion

A hypothetical structure for the major organic constituents of the
Wilcox (Texas) lignite is shown in Figure 2. This is consistent with
experimental data presented.

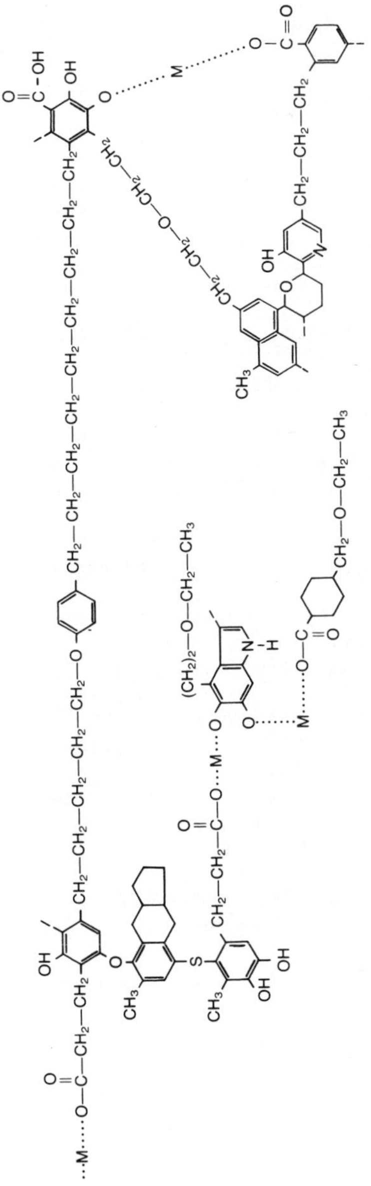

Figure 2.   A Hypothetical Structure for the Organic Constitution of Wilcox (Texas) Lignite.

Acknowledgments

We wish to acknowledge the financial support given by the Gas
Research Institute, the Center for Energy and Mineral Resources of
Texas A&M University and the Robert A. Welch Foundation of Houston,
Texas. Also, we express our appreciation to Dr. B. Hawkins and Dr.
J. Frye of the Colorado State University NMR Center for measuring the
^{13}C CP/MAS NMR spectra. Also, we express our appreciation to Ms.
Argentina Vindiola for invaluable technical assistance.

Literature Cited

1. White, D.M., Kaiser, W.R. and Groat, C.G., Proc 11th Biennial
    Lignite Symp., 1981, p. 112.
2. Bishop, M. and Ward, D.L., Fuel, 1958, 37, 191.
3. "Recommendation   R1994,"   International   Organization   for
    Standardization, 1971.
4. Mallya, N. Ph.D. Dissertation, Texas A&M University, College
    Station, Texas, 1984.
5. Mallya, N. and Zingaro, R.A., "DMSO-A Solvent for the Study of
    Coals" manuscript in preparation.
6. Schafer, H.N.S., Fuel, 1970, 49, 197; ibid., 1970, 49, 271.
7. Teo, K.C., Finora, S. and Leja, J., Fuel, 1982, 61, 17.
8. Deno, N.C., Greigger, B.A. and Stroud, S.G., Fuel, 1978, 57,
    455.
9. Shadle, L. and Given, P.H., personal communication.
10. Philip, C.V. and Anthony, R.G., Fuel Processing Technology,
    1980, 3, 285.
11. Zingaro, R.A., Philip, C.V., Anthony, R.G. and Vindiola, A.,
    Fuel Processing Technology, 1981, 4, 169.
12. Brown, J.K. and Ladner, W.R., Fuel, 1960, 39, 87.
13. Ladner, W.R., Martin, T.G. and Snape, C.E., Preprints ACS Div.
    of Fuel Chem., Vol. 25, No. 4, August 1980.
14. Elliott, M.A., "Chemistry of Coal Utilization," Wiley-Inter-
    science: New York, 1981, p. 486.
15. van Krevelen, D.W. and Schuyer, J., "Coal Science," Elsevier
    Publishing Company, Amsterdam, 1957, p. 145.
16. Williams, R.B. and Chamberlain, N.F., Proc. World Pet. Congr.,
    1964, Sect. V, p. 217.
17. Hippo, E.J. and Walker, P.L., Jr., Fuel, 1975, 54, 245.
18. Jenkins, R.G., Nandi, S.P. and Walker, P.L., Jr., Fuel, 1973,
    52, 288.
19. Morgan, M.E., Jenkins, R.G. and Walker, P.L., Jr., Fuel, 1981,
    60, 189.
20. Walker, P.L., Fuel, 1983, 62, 140.
21. McKee, D.W., Fuel, 1983, 62, 170.

RECEIVED February 17, 1984

Comparison of Hydrocarbons Extracted from Seven Coals by Capillary Gas Chromatography and Gas Chromatography–Mass Spectrometry

SYLVIA A. FARNUM, RONALD C. TIMPE, DAVID J. MILLER, and BRUCE W. FARNUM

Energy Research Center, University of North Dakota, Grand Forks, ND 58202

Seven coals ranging in rank from lignite to bituminous were solvent-extracted with chloroform. The hexane-soluble portions of the extract were analyzed by capillary GC and GC/MS. Selected ion profiles (mass chromatograms) derived from the total ion data were prepared. The coal extract profiles fell into four groups: 1) Big Brown 1, Big Brown 2 (Texas lignites) and Morwell (Australian brown coal); 2) Beulah 3 (North Dakota lignite) and Wyodak (Wyoming, subbituminous); 3) Highvale (Alberta, subbituminous); and 4) Powhatan (Ohio, bituminous). Several groups of hydrocarbon biological markers were identified in the coal extracts including n-alkanes, pristane, sesquiterpenes, several tricyclic alkanes and pentacyclic triterpanes with molecular weights from 398 to 454. The n-alkanes recovered by extraction amount to less than 10% of those produced during liquefaction indicating that bond-breaking reactions during liquefaction lead to the formation of n-alkanes.

Coals have been shown to contain paraffinic hydrocarbons whose structures arise directly from biological precursors (1, 2). The types of hydrocarbons usually considered in this group of biological markers include n-alkanes, acyclic isoprenoids, steranes and terpanes. The amounts of these materials that can be extracted from coal are small but even the characterization of these small amounts of alkanes can be very important in determining the origin and maturity of coals. A detailed understanding of the origin of coals and of the diagenesis and maturation of coals may someday be deduced by drawing upon the large body of available information on the relationships of biological markers to the geologic history of rocks, sediments, oil shales and petroleum. The subject has been reviewed (1,2). Some biological markers have also been detected in coal liquefaction products (3,4).

Experimental

Seven coals were selected for analysis in this study. These seven coals have also been utilized for various liquefaction studies at the University of North Dakota Energy Research Center (UNDERC). Their proximate and ultimate analyses are shown in Table I.

0097–6156/84/0264–0145$06.00/0
© 1984 American Chemical Society

Table I. Analysis of Coals

	Beulah 3 (North Dakota) B3	Morwell (Australia) MOC	Big Brown 1 (Texas) BB1	Big Brown 2 (Texas) BB2	Wyodak (Wyoming) WYO1	Highvale (Alberta) ALB1	Powhatan (Ohio) POW1
Proximate Analysis **As received:**							
Moisture	28.84	10.19	27.55	28.87	30.99	17.79	4.00
Volatile Matter	28.99	43.00	32.10	38.27	30.43	39.83	39.47
Fixed Carbon	30.76	43.63	30.72	19.53	32.96	32.01	46.71
Ash	11.70	3.19	9.63	13.33	5.62	10.37	9.81
Ultimate Analysis **Moisture-free:**							
Ash	16.44	3.55	13.29	18.74	8.15	12.61	10.22
Moisture and ash-free (maf):							
Carbon	69.49	70.64	73.15	74.00	73.74	74.41	79.01
Hydrogen	4.43	5.01	5.22	6.09	5.38	4.92	5.43
Nitrogen	0.99	0.46	1.40	1.22	1.22	0.95	1.29
Sulfur	2.81	0.33	1.30	1.21	0.53	0.20	3.92
Oxygen (by difference)	22.26	23.56	18.93	17.48	19.12	19.52	10.35

The finely pulverized (-100 mesh) coals were sequentially extracted with chloroform (Soxhlet, 6 hr) and the ternary azeotrope chloroform: acetone: methanol, 47:30:23 (Soxhlet, 14 hr). Each extract was separated into a hexane-soluble and a hexane-insoluble portion (Figure 1). The results are shown in Table II. Each of these fractions contained some long chain waxes, tentatively identified as long chain alcohol esters of long chain fatty acids. The only fraction that contained the hydrocarbons of interest was the chloroform soluble-hexane soluble fraction (CHX). The CHX fraction was analyzed by gas chromatography to yield the characteristic profile for the coal. Excellent results were obtained using a J&W DB5 60M fused silica capillary column with H_2 carrier gas and a flame ionization detector. Temperature programming from 50° to 125°C at 0.5°C/min, from 125°C to 250°C at 1.0°C/min, and from 250° to 350°C at 1.5°C/min, followed by an isothermal plateau at 350°C was used. Thermogravimetric analyses showed that >90% of the extracts were volatile.

Table II. Sequential Extraction of Coals with $CHCl_3$
(Soxhlet), $CHCl_3$: Acetone: Methanol, 47:30:23,
Azeotrope (Soxhlet) and Separation of Hexane Solubles
(Wt% maf Coal, Duplicates were Averaged)

	Total $ChCl_3$ Soluble (Soxhlet)	Hexane Soluble Portion of $ChCl_3$ Extract	Total Azeotrope Soluble (Soxhlet)	Hexane Soluble Portion of Azeotrope Extract
B3	2.8	1.0	1.6	0.13
MOC[a]	0.82	0.58	1.7	0.59
BB1[a]	4.2	1.7	2.2	0.16
BB2	3.7	1.7	2.4	0.19
WYO1	3.2	2.9	5.0	0.21
ALB1	0.58	0.35	1.8	0.28
POW1	0.70	0.51	5.8	1.1

[a]Values given from single extraction only.

Results and Discussion

Acyclic Hydrocarbons. Capillary GC comparison of retention times with authentic standards and GC/MS were used to characterize the acyclic alkanes present in the CHX extracts. The extracts of Beulah 3 North Dakota lignite (B3), Wyodak Wyoming subbituminous coal (WYO1), Highvale Alberta subbituminous coal (ALB1), and Powhatan Ohio bituminous coal (POW1) all contained homologous series of n-alkanes, Table III. The low-rank coal CHX extracts for the limited members of the

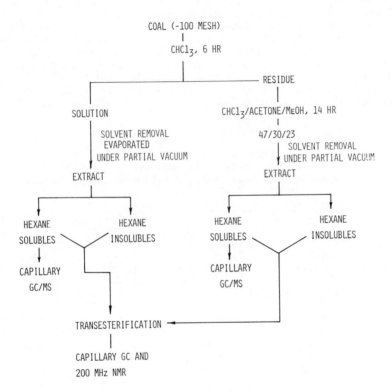

Figure 1.   Sequential Soxhlet extraction of coal.

Table III.   Acyclic Hydrocarbons Found in Extracts of Coals
and Their Distillable Liquefaction Products
(Capillary GC, Area Percent, FID)

	Retention Time, min.	B3 Coal Extract	WY01 Coal Extract	ALB1 Coal Extract	POW1 Coal Extract
Isoprenoids:					
pristane	168.9	0.15			1.35
(2,6,10,14, tetramethyl- pentadecane)					
n-alkanes:					
C-14					
C-15	125.8	0.82			
C-16					
C-17					
C-18					
C-19	194.1	0.74			0.49
C-20	205.0	0.99			0.41
C-21	215.6	1.38			0.26
C-22	223.9	1.22	1.04		0.46
C-23	232.7	0.69	0.73	1.38	0.31
C-24	241.2	0.67	0.33	0.43	0.46
C-25	249.2	1.89	0.82	1.85	0.34
C-26	256.9	1.55	0.57	0.37	0.29
C-27	264.5	2.92	0.99	0.92	0.33
C-28	271.6	0.42	0.31	0.31	0.25
C-29	278.5	0.76	0.98	0.77	0.25
C-30	285.1	0.57	0.52	0.2	0.25
C-31	290.2	0.29	0.56	0.59	0.10
C-32	296.1				0.02
C-33	300.1				0.03
C-34	304.3				0.001
C-35	308.7			0.19	0.03
C-36	312.0			0.09	0.004
C-37	316.6				0.01
C-38	320.4				0.006
C-39	324.0				0.008

series present all gave Carbon Preference Indices, CPI's, (1), greater than one. The alkane distributions and the CPI values are shown
in Figures 2, 3, and 4. The observed n-alkane distributions were in
agreement with the observation by Rigby et al. that increasing maturity of the coal was attended by a shift in maximum concentration to
shorter chain length alkanes in extracts (5).
    As has been previously reported, high temperature high pressure
coal liquefaction as well as coal maturation lowers the odd/even
carbon preference values to near 1.0 (3). This effect was also noted
when the analyses of processed liquid products from the UNDERC continuous processing liquefaction unit were compared with the CHX coal
fractions extracted under mild conditions. CPI values of 1.0 were
obtained for the processed liquids from B3 and WYO1 low-rank coals
compared with values of 1.8 for the fractions extracted under mild
conditions. These coal extracts should contain alkanes that were
present in the coal, whereas the heat-altered, processed products
also contain alkanes formed from cracking of waxes, other alkanes,
and other aliphatic portions of the coal. In the case of the higher
rank bituminous coal, POW1, the CPI was the same, 0.9 both before and
after processing indicating maturation of the coal. Comparison of
the n-alkane extract distributions with the n-alkane continuous
processing unit liquefaction products are also shown in Figures 2, 3,
and 4. If results are calculated as wt % maf coal, the results show
that there are approximately 5 to 10 times larger quantities of
n-alkanes found in the liquefaction products than can be extracted
from the feed coals (Table IV). The same trend was noted by Baset,
who reported 4 wt % maf coal alkanes extracted from Wyodak coal and
2.4 wt % maf coal alkanes obtained by pyrolysis of the coal (6).
Chaffee, Perry, and Johns reported >70% of alkanes formed from pyrolysis of Australian brown coal were non-extractable (7).

Table IV. Extracted/Separated n-Alkanes (Wt % maf Coal)

Coal	Hexane Soluble CHCl$_3$ Extract	Liquefaction Product After 13 Recycle Passes
B3	0.15	1.7
MOC	ND	NA
BB1	ND	3.8
BB2	ND	NA
WYO1	0.20	1.7
ALB1	0.025	NA
POW1	0.022	0.96

ND = Not Detected
NA = Comparable values not available

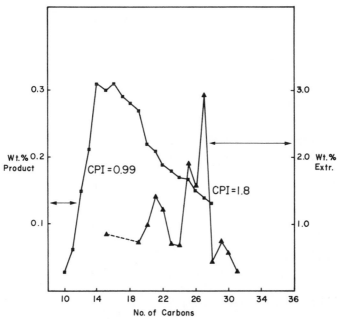

Figure 2. Comparison on n-alkane distribution from Beulah lignite, ▲, and n-alkane distribution from CPU bottoms recycle product, ■ (Beulah lignite feed coal).

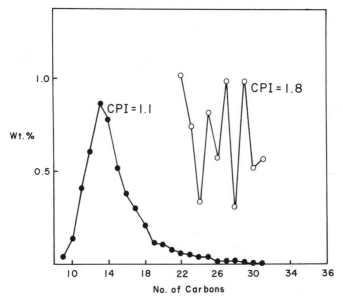

Figure 3. Comparison of n-alkane extract distribution for Wyodak subbituminous coal, O, and n-alkane distribution from CPU bottoms recycle product, ● (Wyodak subbituminous coal).

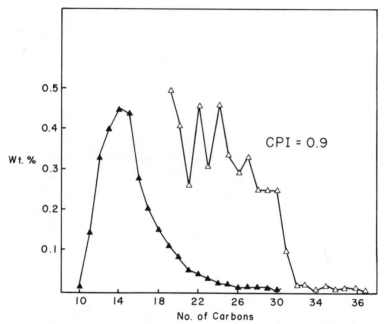

Figure 4.   Comparison of n–alkane extract distribution for
Powhatan bituminous coal, Δ, and n–alkane distribution from
CPU bottoms recycle product (Powhatan bituminous feed coal) ▲.

The CHX extracts from B3 and POW1 were the only extracts of the seven that contained pristane. The bituminous coal, POW1, extract had more pristane than the B3 extract, Table III. Phytane was not detected in any of the extracts, although both pristane and phytane were found in all of the coal liquefaction products from B3, BB1, WYO1, and POW1. The ratio of pristane to phytane was about 5:1. It has been reported that the pristane content of the saturated hydrocarbon fraction from subbituminous coals of more than 76% C begins to increase (1). The increase in phytane begins roughly at 83-85% C.

Cyclic Hydrocarbons. Some of the cyclic hydrocarbons found in the CHX coal extracts are shown in Table V. They were tentatively identified by capillary GC/MS except for the naphthalenes, for which authentic standards were available. Selected ion profiles were used to detect sesquiterpenes (m/e 206, 191), sesquiterpanes (m/e 208), alkanes (m/e 141), alkyl benzenes (m/e 191, 163), steranes (m/e 217), and tricyclic terpenoids (m/e 191, 163) as well as the m/e values for the parent ion of specific compounds. Figure 5 shows the results of one of these selected ion profiles at m/e 206, from the CHX extract of WYO1 along with a portion of the total ion chromatogram. The peaks shown in the m/e 206 trace between scan numbers 240 and 300 correspond to the sesquiterpenes listed in Table V. All of these compounds showed M-15 peaks of various intensities (m/e 191), relatively intense M-29 peaks (m/e 177) and m/e 121 peaks. One compound, c, had a prominent even mass peak at 178. Six had more prominent M-43 (m/e 163) peaks than the others. The fragmentation patterns shown by these $C_{15}H_{26}$ sesquiterpenes resemble those reported by Richardson (8) for a series of $C_{15}H_{28}$ bicyclics identified in a crude oil. We tentatively assigned a C-10 bicyclic structure with one double bond and varying alkyl substituents in accord with the observed fragmentation patterns. Baset et al. assigned similar structures to nine m/e 206 compounds they observed in Wyodak extracts (6). Gallegos (9) also found a number of sesquiterpenes (m/e 206, 191) in pyrolysis products of six coals including Wyodak subbituminous coal and Noonan North Dakota lignite.

The sesquiterpene distribution for the CHX extracts of B3 and WYO1 were distinctive and similar. The distribution for ALB1 gave smaller concentrations of sesquiterpenes, Table V, but the mass spectra were identical for comparable ALB1, WYO1, and B3 sesquiterpenes with the same GC retention times. POW1, MOC, BB1, and BB2 CHX extracts did not contain these compounds. No sesquiterpanes with (m/e 208) were detected in any of the CHX extracts.

Gallegos also detected cadalene (m/e 198, 183) in some of the coal pyrolysis products he investigated (9). Cadalene was present in the ALB1 extract and was probably a small component in the B3 extract. It was not found in the extracts of the other coals.

Another compound which gave identical mass spectra in two of the extracts, WYO1 and ALB1, appears to be one of the tricyclic alkanes, m/e = 276. An intense peak at m/e 247 corresponds to M-29. The mass spectrum was an excellent match with that presented by Philip, et al. (10) for which the structure shown below was proposed. An intense peak at m/e 123 was noted and is probably due to the fragmentation shown.

Table V. Other Hydrocarbons Found in Extracts of Coal, Capillary GC and GC/MS, Area %, FID

	Retention Time, Min.	m/e, Parent Ion	Capillary GC, Area %						
			B3	MOC	BB1	BB2	WY01	ALB1	POW1
Miscellaneous:									
naphthalene	47.9			0.32			0.18	0.12	1.2
2-methylnaphthalene	72.0			0.29			0.19	0.38	2.7
1-methylnaphthalene	75.6			0.04			0.09	0.13	2.0
C_2-naphthalene	95.3								0.68
cadalene	160.4	198	0.31					1.3	1.3
Alicyclic terpenoids:									
$C_{15}H_{26}$ sesquiterpene a	100.7	206					2.8		
" b	104.8	206					1.0		
" c	106.8	206	1.9				13.2	0.70	
" d	108.9	206	1.12				3.3		
" e	111.2	206	0.24				8.3	0.17	
" f	113.0	206					4.6	0.27	
" g	114.9	206	2.3				1.5	1.9	
" h	116.2	206	0.10				0.96	0.17	
$C_{20}H_{36}$ tricyclic alkane	230.3	276	1.7				0.96	0.46	
?	233.4	252						0.50	
$C_{17}H_{30}$ tricyclic alkene	236.4	234	1.2				1.0	0.10	0.10
?	272.5	Not Shown				0.66			
$C_{29}H_{50}$	284.6	298						0.27	0.13
$C_{30}H_{52}$	285.2	412							0.13
$C_{31}H_{54}$	287.0	426				3.3			
$C_{31}H_{54}$	295.8	426							
$C_{32}H_{56}$	312.0	440		0.29	3.0	1.1			
$C_{33}H_{58}$	321.7	454	0.07	0.10	1.7	3.5			0.03

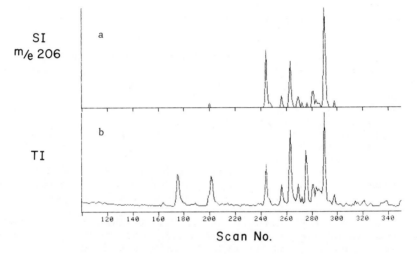

Figure 5. The m/e 206 (a) and TIC (b) chromatograms of
Wyodak subbituminous coal (CHCl$_3$/hexane soluble extract).

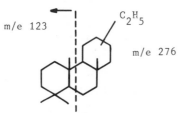

        At a slightly longer retention time the component m/e 234 ap-
pears to be another tricyclic alkane, $C_{17}H_{30}$. Several similar tricyclic alkanes were reported by Jones et al. (4).
        The spectrum of m/e 252, identical in WYO1 and ALB1 extracts,
corresponds to $C_{18}H_{36}$ but is not dodecylcyclohexane (11). There is a
prominent M-15 peak at m/e 237 and relatively intense peaks at m/e
111 and 195. The peak at 111 could be assigned as a dimethylcy-
clohexyl fragment.
        The compounds detected in the mass range 398-454 in B3, BB1,
BB2, and POW extracts appear to be pentacyclic triterpanes since all
have prominent m/e 149, 177, 191, and 205 peaks (12). When selected
ion profiles are compared, WYO1 extract shows no m/e 191 peaks for GC
retention times in the pentacyclic triterpane region. The BB coal
extracts, B3, ALB1 and POW1 all have numerous small peaks that are
probably triterpenoids. Some of these gave appreciable GC FID res-
ponses and good mass spectra.
        The spectrum of m/e 398 is identical for ALB1 and POW1 extracts.
The compound is not adiantane since the intensity of the m/e 191 peak
is greater than 177, not similar (13). C_{27} - C_{31} Triterpanes, especially adiantane, C_{29}, have been found in bituminous coal extracts
(14).
        The pentacyclic component at m/e 412 resembles the spectrum of
oleanane (structure shown) or gammacerane. It is also similar to
that of lupane (13, 15) but lacks the M-43 (369) fragment indicative
of any isopropyl sidechain. Therefore, ring E is probably 6-member-
ed, not 5-membered as in lupane or the hopanes. Although the mass
spectrum is a slightly better match to gammacerane, the short GC
retention time probably rules out gammacerane but not oleanane (13).

$C_{30}H_{52}$ m/e 412

        The two different $C_{31}H_{54}$ pentacycles (m/e 426), one observed in
BB2 extract and one in POW1 extract had no M-29, M-43 or M-57 fragments. The m/e 205 was small, ruling out structures like lupan-3-one
(16). Other differences also eliminate friedelan-3-one and oleanan-
3-one type of structure (16).
        The m/e 440 pentacyclic representative found in BB1 and BB2 had
an M-29 peak (411) and a prominent peak at 369. It could therefore
be a homohopane. None of pentacyclic structures found in any of the
extracts could correspond to the unaltered, less stable 17βH, 21βH

hopane series because m/e 191 is much larger than the secondary fragment in all cases (17). The structure could be a 17αH, 21βH hopane which is common in geologically altered sediments of all kinds.

$$CH_3$$
$$CH$$
$$(CH_2)_n CH_3$$
$$n = 0,1,2...$$

    The m/e 454 pentacyclic triterpene (BB1, BB2 extracts) is also unidentified.
    A number of hydrocarbon fractions from liquefaction products produced at GFETC from the coals studied here were investigated using selected ion profiles of the capillary GC/MS data. None of the hydrocarbon biological markers other than the acyclic alkanes already discussed were detected. There were no sesquiterpenes, sesquiter- panes, steranes, triterpanes or other cyclic terpenoids found.

GC Profiling of Extracts. There were some striking similarities and differences between profiles for the coal CHX extracts. The profiles for CHX extracts of MOC, BB1 and BB2 were very similar. They show no detectable alkanes and very few peaks until 230 min when the maximum temperature was reached. Between 230 min. and 320 min. a large group of peaks form a dense envelope.
    The capillary GC profiles for B3 and WY01 extracts are also strikingly similar, with nearly every major GC peak having a counter- part in each chromatogram. The ALB1 extract GC trace shows some similarity to the WY01 trace but differs, the POW1 extract profile is also different. All of the GC profiles for B3, WY01, ALB1 and POW1 CHX extracts have continuous baseline resolved peak distributions from about 50 to 350 minutes with various recognizable patterns. From duplicate extractions of the same coal and by comparing BB1 and BB2 extracts, it may be seen that the hydrocarbon extracts give recogniz- able profiles that may be used to identify and group coals.

Summary

The chloroform extraction of seven coals yielded extracts that gave unique capillary GC profiles. The coal extract profiles fell into four groups: 1) BB1, BB2 Texas lignites) and MOC (Australian lig- nite); 2) B3 (North Dakota lignite) and WY01 (Wyoming, subbitumin- ous); 3) ALB1 (Alberta, Canadian subbituminous); and 4) POW1 (Ohio, bituminous coal).
    Several groups of hydrocarbon biological markers were identified in the coal extracts by mass spectrometry using selected ion scans. n-Alkanes, pristane, sesquiterpenes, several tricyclic alkanes and pentacyclic triterpanes with molecular weights from 398 to 454 were detected.
    Pristane percentages increase with rank; lignite contained small or non-detectable amounts. There was no apparent correlation between wt % n-alkanes extracted and rank. The distribution maximum chain

length decreases slightly with rank and CPI values decreased with
rank. Over 90% of n-alkanes produced during liquefaction or pyrolysis are non-extractable, leading to the supposition that bond-breaking reactions during liquefaction lead to the formation of n-alkanes.

Acknowledgments

The authors would like to thank Art Ruud for the extensive coal
analyses he and his group carried out. Thanks also to John Diehl for
running the capillary GC profiles of the seven coal extracts.

Literature Cited

1. Bartle, K.D., Jones, D.W., Pakdel, H. "Analytical Methods for
    Coal and Coal Products," Academic Press, New York, San Fran-
    cisco, London, 1978; Part II, 209-262.
2. Bartle, K.D., Jones, D.W., Pakdel, H. "Coal and Coal Products:
    Analytical Characterization Techniques," American Chem. Society
    Symp. Ser., American Chemical Society, Washington, D.C., Volume
    205, 1982, 27-45.
3. White, C.M., Shultz, J.L., Sharkey, A.G., Jr. Nature, 1977,
    268, 620-22.
4. Jones, D.W., Pakdel, H., and Bartle, K.D. Fuel, 1982, 61,
    44-52.
5. Rigby, D., Batts, B.D., and Smith, J.W. Org. Geochem., 1981, 3,
    29-36.
6. Baset, Z.H., Pancirov, R.J., and Ashe, T.R. Adv. Org. Geochem.,
    Proc. Int. Mtg., 9th, 1979, 619-30.
7. Chaffee, A.L., Perry, G.J., and Johnson, R.B. Fuel, 1983, 62,
    311-16.
8. Richardson, J.S. and Miller, D.E. Anal. Chem., 1982, 54, 765-
    768.
9. Gallegos, E.J. J. Chromatog. Sci., 1981, 19, 156-160.
10. Philip, R.P, Gilbert, T.D., and Friedrich, J. Geochim.
    Cosmochim. Acta, 1981, 45, 1173-80.
11. Rubinstein, I. and Strausz, O.P. Geochim. Cosmochim. Acta,
    1979, 1387-92.
12. Kimble, B.J., Maxwell, J.R., Philip, R.P, Eglinton, G., Al-
    brecht, P., Ensminger, A., Arpino, P., and Ourisson, G.
    Geochem. Cosmichim. Acta, 1974, 38, 1165-81.
13. Henderson, W., Wollrab, V., Eglinton, G. Adv. Org. Geochem.,
    Proc. Int. Mtg., 4th, 1968, (1969), 181-207.
14. Allan, J., Bjorory, M., Douglas, A.G. Adv. Org. Geochem.,
    Proc. Inst. Meet., 7th, 1975, 633-54.
15. Ekweozor, C.M., Okogun, J.I., Ekong, D.E.U., and Maxwell, J.R.
    Chem. Geol. 1979, 27, 11-28.
16. Budzikiewicz, H., Wilson, J.M., Djerassi, C.D. J. Am. Chem.
    Soc., 1963, 85, 3688-99.
17. Ensminger, A., Van Dorsselaer, A., Spyckerelle, Ch., Albrecht,
    P., Ourrison, G. Adv. Org. Geochem., Proc. Int. Mtg., 6th,
    1974, 245-60.

RECEIVED February 17, 1984

Analysis of the Inorganic Constituents in Low-Rank Coals

G. P. HUFFMAN and F. E. HUGGINS

U.S. Steel Corporation, Technical Center, Monroeville, PA 15146

Computer-controlled scanning microscopy (CCSEM), Mössbauer spectroscopy and extended X-ray absorption fine structure (EXAFS) spectroscopy have been used to investigate the inorganic constituents of low-rank coals. Mössbauer spectroscopy provides quantitative analysis of all iron-bearing phases, CCSEM yields chemical and size distribution data for discrete mineral phases, and EXAFS allows structural analysis at the atomic level of inorganics dispersed through the macerals. The inorganic phase distribution observed in lignites typically consists of 10-20% quartz, 20-40% kaolinite, 0-30% pyrite, 1-5% of iron sulfates, iron oxyhydroxide, barite and others, and 10-40% of dispersed calcium, with very little calcite or illite ($\lesssim 2\%$). The EXAFS data are consistent with a model in which Ca is bonded to six oxygen ions at an average nearest-neighbor distance of 2.4 Å, has an electronic state similar to that of calcium acetate, and is randomly and molecularly dispersed throughout the coal macerals.

In recent years, numerous modern analytical techniques have been applied to the analysis of the inorganic constituents in coal. At the U. S. Steel Technical Center, Mössbauer spectroscopy and computer-controlled scanning electron microscopy have been emphasized (1-5). With the advent of very intense synchrotron radiation sources, the technique of extended X-ray absorption fine structure (EXAFS) spectroscopy has been applied in many areas of materials science, and several very recent articles on EXAFS studies of the inorganic constituents of coal have appeared (6-8). For the most part, previously published work has reported investigations of bituminous coals by these three techniques. In this chapter, we present some examples of the analysis of the inorganic constituents of lower-rank coals, principally lignites, by these methods. Although the suite of low-rank coals investigated is rather limited, some distinct differences

0097–6156/84/0264–0159$06.00/0
© 1984 American Chemical Society

between the inorganic phase distributions in these coals and those in
bituminous coals are apparent.

Experimental

Mössbauer spectroscopy is a spectroscopy based on the resonant
emission and absorption of low-energy nuclear gamma rays. The ^{57}Fe
nucleus exhibits the best Mössbauer properties of all isotopes for
which the Mössbauer effect has been observed, and ^{57}Fe Mössbauer
spectroscopy is perhaps the best method available for quantitative
analysis of the iron-bearing phases in complex, multiphase samples.
As discussed in recent review articles (2, 9, 10), every iron-bearing
compound exhibits a characteristic Mössbauer absorption spectrum, and
the percentage of the total iron contained in each phase can be
determined from absorption peak areas. Detailed descriptions of the
Mössbauer spectrometer and data analysis programs used in this
laboratory and discussion of the physical basis of the Mössbauer
technique are given elsewhere (10).

The computer-controlled scanning electron microscope (CCSEM)
developed in this laboratory consists of an SEM interfaced by a mini-
computer to a beam-control unit and an energy dispersive X-ray
analysis system. Detailed descriptions of this instrument and its
use in the determination of coal mineralogy and other applications
are given elsewhere (3, 5, 11). Briefly, the electron beam is
stepped across the sample in a coarse grid pattern, with typically
300 x 300 grid points covering the field of view. At each point, the
backscattered electron intensity is measured by detecting those
electrons that are scattered from the sample at angles close to 180°.
When this intensity is greater than a certain preset level, the mini-
computer identifies the feature in question as a mineral particle.
The grid density is then increased to 2048 x 2048 and the particle
area is measured. Next, the beam-control unit brings the beam back
to the geometrical center of the particle and an energy dispersive
X-ray spectrum is collected. Each particle is then placed into one
of up to 30 categories (minerals, compounds, etc.) on the basis of
the chemistry indicated by its X-ray emission spectrum, and approximate weight percentages of all categories are calculated using known
densities and measured particle areas. CCSEM is capable of measuring
the size and chemical composition of up to 1000 particles per hour
for many kinds of particulate samples.

EXAFS spectroscopy examines the oscillatory fine structure above
the absorption edge in the X-ray absorption spectrum of a particular
element. These oscillations arise from interference between the out-
going photoelectron wave and scattered waves produced by interaction
of the photoelectrons with neighboring atoms. As discussed elsewhere
(12, 13), Fourier transform techniques can be used to extract from
these oscillations information about the bond distances, coordination
numbers, and types of ligands surrounding the absorbing element.
Additional information about the valence or electronic state of the
absorbing ion and the ligand symmetry can be obtained from examining

the X-ray absorption near-edge spectra, or XANES, in the energy region very close to the absorption edge (within approximately ±20-30 eV). Such XANES spectra frequently provide characteristic finger-prints for different types of ligand bonding to an absorbing ion (6, 7, 14, 15).

The X-ray absorption spectra of calcium-containing coals and reference compounds discussed in this paper were recorded at the Stanford Synchrotron Radiation Laboratory during a dedicated run of the Stanford Positron-Electron Acceleration Ring at an electron energy of 3.0 GeV. The calcium K-edge occurs at 4038 eV and data were collected from 3800 to 5000 eV, using a double Si (111) monochromator and a fluorescence detector similar to that of Stern and Heald (16). A more detailed discussion of this work is given elsewhere (17).

The samples examined were predominantly lignites from the Pust seam in Montana. However, data for two North Dakota lignites, for slagging and fouling deposits produced by those lignites, and for several subbituminous coals are also included.

Results and Discussion

Mössbauer spectroscopy results for all samples investigated are summarized in Table I. The percentages of the total iron contained in the sample assigned to each of the iron-bearing minerals identi-fied are given in columns 1 to 4, and the weight percentages of pyritic sulfur, determined from the pyrite absorption peak areas as discussed elsewhere (1), are given in column 5. It is seen that pyrite and minerals (iron sulfates and iron oxyhydroxide) that are probably derived from pyrite by weathering are the only iron-bearing species in these low-rank coals. Notably absent are contributions from iron-bearing clays and siderite which are common constituents of bituminous coals (1, 2). Pyrite and iron oxyhydroxide are difficult to separate with room-temperature Mössbauer spectroscopy (18). For example, in Figure 1 are shown the room temperature and 77 K spectra obtained from the Pust seam, C, lignite, which had been stored for several years prior to measurement. Although it is quite difficult to determine the relative amounts of pyrite and oxyhydroxide from the room-temperature Mössbauer spectrum, the spectral contributions of the two phases are readily resolved at 77 K.

CCSEM results for the approximate weight percentages of all inorganic phases are given in Table II. Perhaps the most interesting aspect of the CCSEM results for these low-rank coals as compared with similar data for bituminous coals is the abundance of Ca-rich phases. In most cases, these phases are not calcite, but are Ca-enriched macerals in which the Ca is uniformly dispersed throughout the coal, as illustrated by Figure 2. The Ca-enriched macerals appear light gray in the backscattered electron image shown in Figure 2a; the dark gray material is the epoxy mounting. The relatively uniform dispersion of the Ca is illustrated by the Ca X-ray map of Figure 2b, and an energy dispersive X-ray spectrum obtained from an individual

TABLE I

Mossbauer Results for Low-Rank Coals

| Sample | Percent of Total Iron Contained in | | | | Wt. % of |
	Pyrite	Jarosite	Ferrous Sulfate	Iron Oxy-hydroxide	Pyritic Sulfur
Pust seam, A-3	100	-	-	-	0.30
Pust seam, A-4	92	7	1	-	2.26
Pust seam, A-6	91	9	-	-	0.15
Pust seam, A-7	100+	-	-	-	0.03+
Pust seam, B-3	100	-	-	-	0.06
Pust seam, B-5	100	-	-	-	0.05
Pust seam, B-7	100	-	-	-	0.09
Pust seam, C	43	26	-	31	0.29
N.Dakota lignite, heavy fouling	91	6	2	-	0.50
N.Dakota lignite, light fouling	95	5	-	-	0.38
Rosebud subbituminous	65	16	19	-	0.15
Colstrip subbitumnous	81	19	-	-	0.27

+Very weak spectrum; sample contained only 0.07% iron.

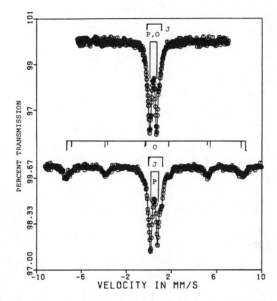

Figure 1.   Mossbauer spectra of the Pust seam, C, lignite.
Pyrite (P), jarosite (J), and iron oxyhydroxide (O) are
indicated.   Key: top, room temperature; and bottom, 77 K.

TABLE II

CCSEM Results for Low-Rank Coals (approximate weight percentages)

Sample	Quartz	Kaolinite	Mixed Silicates	Illite	Pyrite	Fe Sulfates	Fe- rich	Ca- rich+	Barite	Apatite	Other Phases	
Pust seam, A-3	7	45	8	-	1	-	6	26	1	-	Smectite Rutile	1 1
Pust seam, A-4	12	20	3	-	26	4	7	15	7	2	Fe-Ca-S	3
Pust seam, A-6	7	27	-	-	1	-	-	51	5	3	-	
Pust seam, A-7	7	35	3	-	1	-	-	49	3	-	-	
Pust seam, B-3	10	43	13	1	1	-	-	22	3	3	Smectite	4
Pust seam, B-5	14	38	3	-	1	-	-	38	5	-	Rutile	1
Pust seam, B-7	16	38	2	-	5	-	-	27	5	-	Ca-Fe-S	1
Pust seam, C	9	29	5	1	4	5	14	13	4	-	Ca-Fe Ca-Sr-Al-P* Jarosite	6 4 2
N.Dakota lig., heavy foul.	24	13	20	3	19	1	1	14	-	-	Smectite	1
N.Dakota lig., light foul.	28	16	13	2	10	-	2	23	-	-	Smectite	1
Rosebud, Sbb.	23	36	6	2	6	2	3	8	1	1	Ca-Sr-Al-P* Smectite Rutile	6 1 1
Eveleth, Sbb.	18	23	22	12	4	-	3	7	-	1	Ca-Sr-Al-P*	8
Colstrip, Sbb.	22	25	6	3	8	-	3	29	-	-	Gypsum	4
Absaloka, Sbb.	56	22	8	-	3	-	-	5	-	-	Rutile	3

+As discussed in the text, percentages given for the Ca-rich category are not quantitative.

*Minerals denoted by Ca-Sr-Al-P are probably crandallite $(Ca,Sr)Al_3((PO_4)_2(OH)_5 H_2O$.

Figure 2.   Backscattered electron image of Pust lignite (a), Ca X-ray map of the same area (b), and an energy-dispersive X-ray spectrum from an individual maceral (c).

maceral is shown in Figure 2c. As discussed below, EXAFS data indicate that this dispersed calcium is present as salts of carboxylic acids.

The backscattered electron intensity of the Ca-enriched macerals is significantly smaller than that of calcite, and CCSEM can make a distinction, albeit somewhat imprecisely, between Ca-enriched macerals and calcite or other Ca-rich minerals on this basis. However, the CCSEM programs have not been properly calibrated to deal with the case of macerals enriched in an inorganic component such as Ca at this point. Consequently, the percentages indicated in Table II for the Ca-rich category are only a qualitative indication of the relative amounts of this species in the various low-rank coals examined. On the basis of the backscattered electron intensity, it appears that calcium is dispersed throughout the macerals of the lignites that have been examined, and is present partially in dispersed form and partially as calcite in the subbituminous coals. In fresh bituminous coals, calcium is present almost exclusively as calcite (3-5).

For comparison with Tables I and II, Table III gives the range and typical values of the mineral distributions observed in bituminous coals by the CCSEM and Mossbauer techniques, derived from studies of perhaps a hundred different bituminous coal samples in this laboratory. Some obvious differences in mineralogy are apparent. In addition to the difference in calcium dispersion and abundance already noted, it is seen that certain minerals common in bituminous coals, such as Fe-bearing clays (illite and chlorite) and siderite, are virtually absent in the low-rank samples of Tables I and II. Conversely, minerals such as barite ($BaSO_4$), apatite ($Ca_5(PO_4)_3OH$), and other Ca, Sr phosphates are rather uncommon in bituminous coals.

EXAFS and XANES data for a Ca-rich sample of the Pust seam, A, lignite can be briefly summarized by reference to Figures 3 and 4. Figure 3 shows the XANES of the lignite, calcium acetate, and a fresh bituminous coal from the Pittsburgh seam rich in calcite. The strong similarity between the lignite and calcium acetate spectra is apparent. Similarly, a close similarity is also observed for the XANES of the fresh bituminous coal and that of a calcite standard. Examination of the XANES of several other standard compounds (CaO, $Ca(OH)_2$, and $CaSO_4 \cdot 2H_2O$) showed that none of these phases were present in detectable amounts in either coal. The strong similarity of the XANES of calcium acetate and that of the lignite is direct evidence that the calcium in this coal is associated with carboxyl groups in the macerals and is not contained in very fine ($<0.1 \mu m$) mineral matter.

Mathematical analysis of the EXAFS associated with the nearest-neighbor oxygen shell surrounding the Ca^{2+} ions in the lignite was accomplished using programs developed by Sandstrom (13). Briefly, the results indicate that the Ca is coordinated by six oxygens, possibly contributed in part by water molecules, at an average nearest-neighbor distance of 2.39 Å. Additionally, the EXAFS data indicate that structural order at distances further from the Ca^{2+} ions than the first coordination shell is essentially absent in the lignite, implying that the Ca sites are more or less randomly

TABLE III

Mineralogy of Bituminous Coals

CCSEM Analysis, Wt. % of Mineral Matter			Mossbauer Analysis, % of Total Sample Iron		
Mineral	Range	Typ-ical	Mineral	Range	Typ-ical
Quartz	5-44	18	Pyrite	25-100	62
Kaolinite	9-60	32	Ferrous Clay	0-55	18
Illite	2-29	14	Siderite/Ankerite	0-58	9
Chlorite	0-15	2	Ferrous Sulfate	0-18	3
Mixed Silicates	0-31	17	Jarosite	0-21	4
Pyrite	1-27	8	Wt. % Pyritic	0.08-	0.35
Calcite/Dolomite	0-14	3	sulfur*	1.51	
Siderite/Ankerite	0-11	2			
Other	0-12	4			

*Determined by method of Ref. 1.

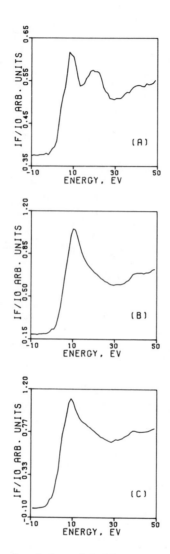

Figure 3. XANES of calcite-rich bituminous coal (A), the Pust seam lignite (B), and calcium acetate (C).

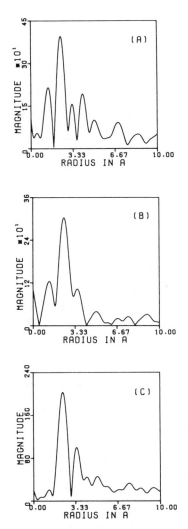

Figure 4.   Magnitudes of the phase-shift subtracted Fourier
transform of calcite-rich bituminous coal (A), the Pust seam
lignite (B), and calcium acetate (C).

distributed throughout the macerals. This point is well illustrated
by phase-shift subtracted Fourier transforms of the EXAFS data,
which, as discussed elsewhere (12, 13), should bear a reasonably
close relationship to the radial distribution functions appropriate
for the local environment of the Ca^{2+} ions. In Figure 4, the
magnitude of the phase-shift subtracted Fourier transform, $|F(r)|$, is
shown for a calcite-rich bituminous coal from the Pittsburgh seam,
the lignite sample, and calcium acetate. It is seen that the $|F(r)|$
curves for calcium acetate and the calcite-rich bituminous coal
exhibit maxima corresponding not only to the nearest-neighbor oxygen
shell, but also at the approximate locations of more distant calcium
neighbor shells. The $|F(r)|$ curve of the lignite exhibits a clear
maximum only at the oxygen nearest-neighbor shell distance. It is
possible, however, that the small shoulder that appears on the high
side of the oxygen shell peak in the lignite $|F(r)|$ curve (at
approximately 3.5 Å) could correspond to the initial stages of Ca^{2+}
ion clustering.

The calcium XANES and the phase-shift subtracted Fourier transforms of the EXAFS obtained from coals of several different ranks
(lignite, subbituminous, high volatile C bituminous, and high volatile A bituminous) are compared in Figures 5 and 6. These spectra
illustrate the progression of the form of calcium from highly dispersed carboxyl group cations in the lignite to well-crystallized
calcite in the hvA bituminous coal. Increases occur in the intensities of certain features in both the XANES (at about 5 and 18 eV)
and the Fourier transforms (at approximately 3.5, 4.0, and 5.5 Å)
with increasing rank. The increase of peak intensities in the
Fourier transforms presumably arises from the formation of ordered
calcium shells at various distances from the central calcium ion.
The changes in the XANES reflect differences in the bound electronic
levels of Ca^{2+} ions in calcite from those of Ca^{2+} ions bonded to
carboxyl groups in the macerals.

It is worth noting that XANES and EXAFS spectra obtained from a
severely weathered bituminous coal were nearly identical to those
obtained from lignite samples. This indicates that in the weathered
coal calcium is also present in a dispersed form in which it is
bonded to carboxyl groups in the macerals. A more detailed report of
this work is given elsewhere (17).

Analysis of Fouling and Slagging Deposits. It was observed that
one of the two North Dakota lignites listed in Tables I and II produced heavy fouling deposits during combustion in a large utility
furnace, while little or no difficulty was experienced in firing the
other. As seen in Tables I and II, the inorganic phase distributions
of these two coals are rather similar. Additionally, CCSEM and
Mössbauer analysis of boiler-wall slag deposits produced by both
coals gave rather similar results. A typical Mössbauer spectrum obtained from a boiler wall deposit (wall temperature ∿1250°C) is shown
in Figure 7, and a summary of the approximate phase distributions of
the wall slag deposits produced by both coals is given in the inset.

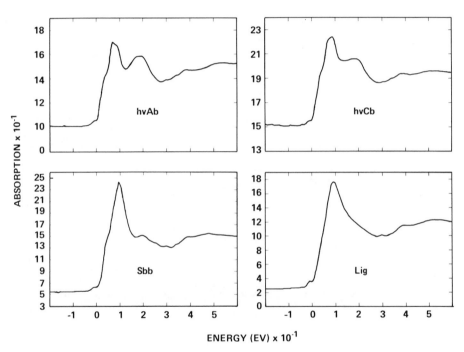

Figure 5.   Calcium XANES spectra obtained from coals of four different ranks.

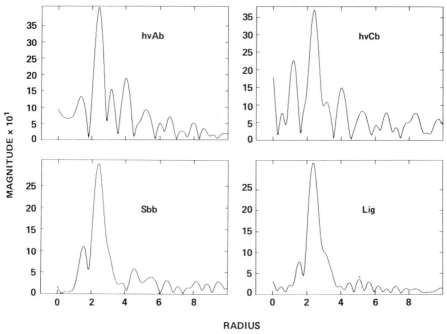

Figure 6. Magnitudes of the phase-shift subtracted Fourier transforms of the EXAFS obtained from coals of four different ranks.

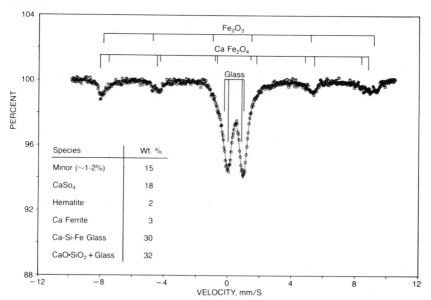

Figure 7. Mossbauer spectrum of boiler-wall slag deposits. The approximate phase distribution determined from CCSEM and Mossbauer results is shown in the inset.

TABLE IV

CCSEM Results for Fouling Deposits, 800-1000 °C

Species	Heavy (wt%)	Light (wt%)
Al, Si-rich	5	5
Quartz	3	8
Hematite	2	3
Ca-rich	10	7
Mg-Ca	3	5
Ca Ferrite	2	<1
Ca-Si-Fe Glass	21	22
Ca-Si + Glass	40	40
Alkali Sulfate	2	0
$CaSO_4$ + Alkali Sulfate	4	$CaSO_4$ only - 6
Other	8	4

Alkali Sulfate - Na/Ca/K - 1.0/0.5/0.1.

From the observed phases, it appears that the $CaO-SiO_2-Fe_2O_3$ phase diagram plays a key role in determining the slagging behavior of these coals. CCSEM analyses of fouling deposits produced by the two lignites are given in Table IV. It is seen that the only significant difference between the light and heavy deposits is the presence of an alkali sulfate mixture, containing Na, K, and Ca sulfates, in the latter. It is well known that alkali sulfates can react strongly with metal surfaces to produce alkali-iron sulfate mixtures that are partially molten over the temperature range from approximately 700 to 1100 °C, causing severe fouling and corrosion problems (19).

Conclusions

In this paper, we have presented some examples of the usefulness of three techniques, Mossbauer spectroscopy, CCSEM, and EXAFS in the analysis of low-rank coals and combustion products of low-rank coals. Points of interest regarding the inorganic constituents of these coals include the high abundance of calcium bonded to carboxyl groups and dispersed throughout the macerals, the low abundance of illite, siderite, and calcite, and the presence of significant amounts of accessory minerals such as barite, apatite, and other phosphates. Areas that merit further investigation by these techniques include the analysis of fouling and slagging deposits, and structural studies of inorganic elements such as calcium that are dispersed through and bonded to the coal macerals.

Acknowledgments

The EXAFS experiments discussed in this paper were conducted in collaboration with F. W. Lytle and R. B. Greegor of the Boeing Company. We are grateful to the Stanford Synchrotron Radiation Laboratory and the supporters of that facility, DOE, NSF, and NIH for beam time, and for assistance from the excellent SSRL staff.

The material in this paper is intended for general information only.
Any use of this material in relation to any specific application
should be based on independent examination and verification of its
unrestricted availability for such use, and a determination of suitability for the application by professionally qualified personnel.
No license under any United States Steel Corporation patents or other
proprietary interest is implied by the publication of this paper.
Those making use of or relying upon the material assume all risks and
liability arising from such use or reliance.

Literature Cited

1. Huffman, G. P.; Huggins, F. E. Fuel 1978, 57, 592–604.
2. Huggins, F. E.; Huffman, G. P. "Anal. Methods for Coal and Coal
    Products," Vol. 3, Clarence Karr, Jr., Ed.; Academic: NY, 1979,
    371–423.
3. Huggins, F. E.; Kosmack, D. A.; Huffman, G. P.; Lee, R. J.
    SEM/1980/I SEM, Inc.: AMF O'Hare (Chicago), IL, 1980, 531–540.
4. Huffman, G. P.; Huggins, F. E.; Lee, R. J. "Proc. Intern. Conf.
    Coal Science, Dusseldorf," Verlag Gluckauf GmbH: Essen, Germany,
    1981, 835–840.
5. Huggins, F. E.; Huffman, G. P.; Lee, R. J. "Coal and Coal
    Products: Analytical Characterization Techniques," E. L.
    Fuller, Jr., Ed.; ACS Symposium Series 205, Amer. Chem. Soc.:
    Washington, DC, 1982, 239–258.
6. Maylotte, D. H.; Wong, J.; St. Peters, R. L.; Lytle, F. W.;
    Greegor, R. B. Science 1981, 214, 554.
7. Sandstrom, D. R.; Filby, R. H.; Lytle, F. W.; Greegor, R. B.
    Fuel 1982, 61, 195.
8. Hussain, Z.; Barton, J. J.; Umbach, E.; Shirley, D. A. "SSRL
    Activity Report," A. Bienenstock and H. Winick, Eds.; May 1981,
    VII-102.
9. Huffman, G. P.; Huggins, F. E. "Mossbauer Spectroscopy and Its
    Chemical Applications," J. G. Stevens and G. K. Shenoy, Eds.;
    Advances in Chemistry Series 194, Amer. Chem. Soc.: Washington,
    DC, 1981, 265–301.
10. Huffman, G. P.; Huggins, F. E. "Physics in the Steel Industry,"
    F. C. Schwerer, Ed.; AIP Conf. Proc. 84, Amer. Inst. Physics:
    NY, 1982, 149–201.
11. Lee, R. J.; Kelly, J. F. SEM/1980/I SEM, Inc.: AMF O'Hare
    (Chicago), IL, 1980, 303–310.
12. Lee, P. A.; Citrin, P. H.; Eisenberger, P.; Kincaid, B. M.
    Rev. Mod. Phys. 1981, 53, 769–806.
13. Sandstrom, D. R. "Proc. Workshop on EXAFS Analysis of Disordered
    Systems, Parma, Italy, 1981," to be published in Nuovo Cimento B
14. Kutzler, F. W.; Natoli, C. R.; Misemer, D. K.; Doniak, S.;
    Hodgson, K. O. J. Chem. Phys. 1980, 73, 3274–88.
15. Bair, R. A.; Goddard, W. A.; Phys. Rev. B 1980, 22, 2767–76.
16. Stern, E. A.; Heald, S. Rev. Sci. Instrum. 1979, 50, 1579.
17. Huggins, F. E.; Huffman, G. P.; Lytle, F. W.; Greegor, R. B.
    "Proc. Intern. Conf. Coal Science, Pittsburgh," Intern. Energy
    Agency: NY, 1983, 679–682.
18. Huggins, F. E.; Huffman, G. P.; Kosmack, D. A.; Lowenhaupt,
    D. E. Intern. J. Coal Geol. 1980, 1, 75–81.
19. Reid, W. T. "Chemistry of Coal Utilization" (Suppl. Vol.), M. A.
    Elliot, Ed.; John Wiley & Sons, Inc.: NY, 1963, 1389–1445.

RECEIVED February 17, 1984

Geochemical Variation of Inorganic Constituents in a North Dakota Lignite

F. R. KARNER, S. A. BENSON, HAROLD H. SCHOBERT, and R. G. ROALDSON

Energy Research Center, University of North Dakota, Grand Forks, ND 58202

A study of the variation in the distribution of major, minor, and trace elements in lignite-bearing lithologic sequences is critical for both interpretation of depositional environments and utilization of the coal. Samples of lignite of the Kinneman Creek Bed, underclay and overburden from the Center Mine were analyzed by neutron activation, x-ray fluorescence, and x-ray diffraction techniques. Major patterns of element distribution in the lignite include (1) concentration in the margins of the seam (Al, Si, S, Sc, Fe, Co, Ni, Zn, As, Zr, Ag, Ba, Ce, Sm, Eu, Yb, Th, and U), (2) concentration in the lower part of the seam (Cl, K, V, Cr, Cu, Ge, Se, Ru, Sb, Cs, and La), and (3) even distribution throughout the seam (Na, Mg, Ca, Sr, and Mn). Other elements exhibit indefinite or irregular patterns (P, Rb, Y, Cd). Elemental variation patterns in the lignite may be related to admixture of detrital material during deposition, secondary precipitation of minerals, and ionic exchange processes involving adsorption of ions from circulating groundwater by both organic and inorganic components.

This chapter summarizes information on the distribution of major, minor, and trace elements in a lithologic sequence of sedimentary materials in a major lignite-producing portion of the Fort Union region of the northern Great Plains. The Kinneman Creek Bed and associated sediments at the Center Mine in the Knife River Basin exhibit several types of spatial patterns of enrichment and depletion of chemical elements. Previous studies of lignite-related geochemical variation in this region are summarized in the literature (1,2). The principal emphasis of this chapter is concerned with the patterns of elemental variations in lignite. These patterns are related to factors of accumulation of both organic and inorganic components during deposition and aqueous precipitation, dissolution, and ion-exchange processes after deposition. Elemental variations in overburden and underclay are dependent upon relative abundances of quartz- and feldspar-rich silt fractions versus clay mineral-rich

0097–6156/84/0264–0175$06.00/0
© 1984 American Chemical Society

fractions of the sediment, clay mineral and carbonate mineral variation, devitrification of volcanic ash, and distribution of authigenic cements and organic matter. The data and results presented in this chapter are part of continuing work on the inorganic constituents of low-rank coals and the factors that affect their distribution.

Geologic Relationships

The Center mine, Oliver County, North Dakota, is in the Fort Union coal region, which contains one of the largest reserves of lignite in the world. The Fort Union region encompasses portions of the north-central United States and south-central Canada, including both sub-bituminous coals of the northern Powder River Basin and lignites of the Williston Basin (2). Important lignite beds of the Fort Union Region occur within the Ludlow, Slope, Bullion Creek, and Sentinel Butte Formations of the Paleocene Fort Union Group.
   The Sentinel Butte Formation has been described as a "lignite-bearing, nonmarine, Paleocene unit whose outcrops are somber gray and brown" (3). Rock types include sandstone, siltstone, claystone, lignite, and limestone. The lignite beds typically vary from less than one meter to 3-5 m in thickness (4). The Oliver-Mercer County district is located in eastern Mercer and northeastern Oliver Counties with the major lignite seam, the Hagel Bed, ranging in thickness from 3-4 m and the overlying seam reported on here, designated by Groenewold and others (4) as the Kinneman Creek Bed, up to 2.5 m thick.
   The lignite of the Kinneman Creek Bed is a black to brownish-black coal that slacks rapidly on exposure to the atmosphere and typically contains dark carbonaceous clay or gray clay partings. The underclay is a gray-green clay with lignite fragments. The overburden is gray fine-grained sediment primarily consisting of silty clays and clayey silts with minor concretionary zones and sands. Logan (6) has described the overburden sequence as the Kinneman Creek interval and has interpreted it to be of lacustrine origin. Generally, the early Cenozoic sediments in this region are believed to have been deposited in a coastal complex of stream channel, flood plain, swamp, lake, and delta environments.

Inorganic Constituents

Inorganic constituents in the Sentinel Butte sediments are present as 1) detrital mineral grains and volcanic glass fragments, 2) authigenic mineral grains and cement, 3) inorganic components of plants, and 4) ions adsorbed by clay, other minerals, and organic material (7-10).
   A summary of minerals observed in the Sentinel Butte Formation is given in Table I (9). Major original detrital constituents include montmorillonite, quartz, plagioclase, alkali feldspar, biotite, chlorite, volcanic glass, and rock fragments. Major minerals formed during deposition and diagenesis by conversion of original detrital constituents consist of montmorillonite, chlorite, and kaolinite. Pyrite, gypsum, hematite, siderite, and possibly calcite formed through post-depositional reducing and oxidizing reactions related to changing conditions of deposition, burial, and groundwater movement

and chemistry. Occurrences and modes of formation of these post-depositional constituents are currently under study.

Table I. Inorganic Constituents in the Sentinel Butte Formation (from Schobert and others, 9)

Constituents	Overburden	Lignite	Underclay
Alkali feldspar	XX	X	X
Augite		X	
Barite		X	
Biotite	X		
Calcite/Dolomite	XX	X	
Siderite	XX		
Chlorite	XX	X	X
Gypsum		XXX	
Hematite		XX	
Hornblende	X		
Illite	XX	X	X
Kaolinite	XX	XXX	XXX
Magnetite	X		
Montmorillonite	XXX	X	XXX
Muscovite	X		
Plagioclase	XXX	X	X
Pyrite		XX	
Quartz	XXX	XXX	XXX
Volcanic Glass	X		
Rock fragments	XX		

XXX = Abundant
 XX = Common
  X = Minor

Sampling

The samples used in this study are a subset of those collected in a sampling program (5) which was designed to obtain stratigraphically-controlled specimens for a study of the character, distribution, and origin of the inorganic constituents in lignite. The objectives of the field sampling were to obtain:

1. Incremental samples of underclay, lignite, and overburden;

2. Duplicate vertical sections and lateral samples; and

3. Samples of specific beds, lenses, fault zones, mineral concentrations, or other materials of unusual aspect.

Lignite, lignite overburden, and underclay were collected from a vertical section of a freshly exposed high wall at an actively mined pit in the Center mine. Figure 1 illustrates the lithologic sequence.

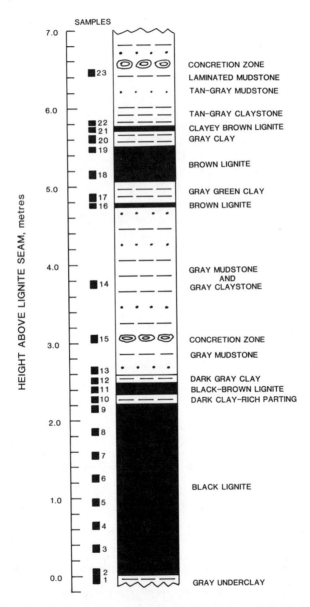

Figure 1. Geologic column and sample locations for the Kinneman
Creek Bed lithologic sequence at the Center Mine sampling section 1–D.

To minimize contamination, material from slope wash and mining activity was removed first. In order to obtain samples of underclay, it was necessary to dig down at the base of the lignite. Once the section to be sampled was exposed, it was measured and marked at 30 cm intervals through the lignite and then at increments of one meter into the overburden. Samples were collected over a 10 cm high by 50 cm wide area at the bottom of each marked interval in homogenous materials. In non-homogenous materials individual layers, lenses, contacts, and other specific features were sampled over a 10 cm by 50 cm area or smaller, as controlled by the dimensions of the sampled unit. All samples were stored in plastic bags inside cardboard cartons. Samples analyzed in this study are listed in stratigraphic sequence in Table II.

Table II. Descriptions and Locations of Samples from Kinneman Creek Bed and Related Sediments, Center Mine, ND

Sample Number	Height in Meters*	Sample Description
1-23-D	6.43	Tan-gray laminated mudstone
1-22-D	5.81	Tan-gray claystone
1-21-D	5.73	Brown lignite with clayey layers
1-20-D	5.60	Gray clay
1-19-D	5.48	Brown lignite, top of seam
1-18-D	5.15	Brown lignite, soft, clayey
1-17-D	4.85	Gray-green clay
1-16-D	4.75	Brown lignite, soft, clayey
1-14-D	3.75	Gray claystone
1-15-D	3.04	Concretion zone
1-13-D	2.65	Gray mudstone
1-12-D	2.51	Dark gray organic-rich clay
1-11-D	2.40	Black-brown lignite, soft, top of seam
1-10-D	2.26	Dark clay-rich parting
1-9-D	2.15	Black lignite, hard
1-8-D	1.85	Black lignite, hard
1-7-D	1.55	Black lignite, hard
1-6-D	1.25	Black lignite, hard
1-5-D	0.95	Black lignite, hard
1-4-D	0.65	Black lignite, hard
1-3-D	0.35	Black lignite, hard
1-2-D	0.05	Black lignite, hard
1-1-D	-0.05	Gray underclay

* Height from base of major lignite to center of sampled interval (10 cm thick except full thickness for thinner units).

Analytical Methods

Samples of coal, overburden, and underclay were analyzed by neutron activation (NAA) and x-ray fluorescence analysis (XRF). NAA analysis

was performed by the Nuclear Engineering Department, North Carolina
State University. The system description, detection limits, and
sample preparation procedures for coal and coal fly ash are summarized by Weaver (11). XRF was done with an energy dispersive x-ray
system (12). Mineralogy of the overburden and underclay was determined by x-ray diffraction (XRD). The crystalline phases present in
the coal were determined by XRD of the low-temperature ash of the
coal. Low temperature ashing of raw coal was done at ≅150°C with 2
torr pressure in an oxygen plasma.

Results and Discussion

Mineralogy of the Lithologic Sequence. The major and minor minerals
present in the overburden, underclay, and low-temperature ash of the
coal are summarized in Table III. The most abundant minerals in the
clay- and silt-rich sediments of the overburden include quartz,
"mica/illite" (a mixture of mica and illite in varying proportions),
plagioclase, kaolinite, and montmorillonite. The mica/ illite group,
recognized by routine x-ray diffraction, is being studied further by
specialized x-ray diffraction, optical, scanning electron microscope
and microprobe methods.
          Underclay and overburden sediments are typically clays with
about 80% clay and 20% silt and muds with about 50% clay and 50%
silt. The most distinct chemical variation in the overburden is the
compact siderite- and dolomite-cemented concretion zone, 4.5 m above
the base of the Kinneman Creek lignite.
          The bulk mineralogy of the coal, also represented in Table III,
is characterized by quartz, calcite, bassanite, kaolinite, and pyrite.
Bassanite is probably a low-temperature ashing product of organic
sulfur fixation with organically bound calcium (13). Pyrite content
appears to increase with depth in the seam relative to calcite and
kaolinite.

Variations of Elements in the Lithologic Sequence. The results of
the neutron activation and x-ray fluorescence analysis for the stratigraphic sequence are listed in Table IV.

Variation Within Major Coal Seam. The distribution of elements
within the coal seam can be summarized by four general trends:
1) concentration of elements in the margins, 2) concentration in the
lower part of the seam, 3) even throughout the seam or minor distribution patterns, and 4) irregular distribution.
          The elements concentrated in the margins of the seam form the
largest group and include Al, Si, S, Sc, Fe, Co, Ni, Zn, As, Rb, Y,
Zr, Ag, Ba, Ce, Sm, Eu, Yb, Th, and U. Figures 2 and 3 illustrate
the distributions of Al and Ni within the lithologic sequence including underclay, the Kinneman Creek coal bed with a clay parting at
2.26 m, and the overburden.
          The elements concentrated in the lower part of the seam include
Cl, K, Ti, V, Cr, Cu, Ge, Se, Ru, Sb, Cs, and La. Figure 4 illustrates the distribution of V.
          The elements which show an even distribution and in some cases
show a slight concentration at the center of the seam in the Kinneman

Table III. Major and Minor Minerals Determined by X-ray Diffraction
Analysis of Sediments and Low-Temperature Ash of Coal

Sample	Height (Meters)	Description	Observed Minerals Major	Minor
1-1-D	-0.05	Gray under-clay	Quartz, kaolinite mica-illite	Plagioclase
1-3-D	0.35	Black lignite	Quartz, gypsum, bassanite, kaolinite	Pyrite, calcite (?)
1-7-D	1.55	Black lignite	Quartz, kaolinite, bassanite	Pyrite, calcite (?)
1-10-D	2.26	Dark clay parting	Quartz, kaolinite, mica-illite	Pyrite, plagioclase
1-11-D	2.40	Black-brown lignite	Quartz, kaolinite, bassanite	Pyrite, calcite (?)
1-12-D	2.51	Dark gray clay	Quartz, mica-illite, kaolinite	Plagioclase
1-13-D	2.65	Gray mudstone	Quartz, mica-illite, kaolinite	Plagioclase
1-14-D	3.75	Gray clay-stone	Quartz, mica-illite, kaolinite	Plagioclase
1-15-D	3.04	Concretion zone	Siderite, dolomite, quartz	Mica-illite, kaolinite, plagioclase
1-17-D	4.85	Gray-green clay	Quartz, mica-illite, kaolinite	Plagioclase
1-20-D	5.60	Gray clay	Quartz, mica-illite, kaolinite	Plagioclase
1-22-D	5.81	Tan-gray claystone	Quartz, mica-illite, kaolinite	Plagioclase
1-23-D	6.43	Laminated mudstone	Quartz, mica-illite, calcite	Plagioclase, montmoril-lonite(?)

Table IV. Elemental Analysis for Selected Major

ID		Height* M.	Ash%~	Na	Mg	Al	Si**	P**
1-1-D	Underclay	-0.05	--	5332	9534	95268	299600	0
1-2-D	Lignite	0.05	14.2	1518	1736	10200	14220	994
1-3-D	Lignite	0.35	11.2	1307	1607	5586	9520	1008
1-4-D	Lignite	0.65	7.6	1398	1677	2595	2270	760
1-5-D	Lignite	0.95	6.3	1398	1375	3313	3150	630
1-6-D	Lignite	1.25	6.6	967	1861	2949	1790	594
1-7-D	Lignite	1.55	9.0	1115	2209	6051	7880	1080
1-8-D	Lignite	1.85	5.3	804	827	1970	858	636
1-9-D	Lignite	2.15	14.3	1006	1979	10813	14700	1144
1-10-D	Clay parting	2.26	--	3685	9219	100370	292100	0
1-11-D	Lignite	2.40	9.7	737	1848	6620	8670	0
1-12-D	Clay	2.51	--	5814	10564	78761	287900	0
1-13-D	Mudstone	2.65	--	5331	10921	71239	291700	0
1-14-D	Claystone	3.75	--	4461	11465	95029	298200	0
1-15-D	Concretion Zone	3.04	--	1422	3673	96920	98600	0
1-16-D	Lignite	4.75	16.2	750	1762	5813	9850	1458
1-17-D	Clay	4.85	--	5767	10951	96097	299600	0
1-18-D	Lignite	5.15	7.5	718	1652	2512	4150	825
1-19-D	Lignite	5.48	9.6	609	2208	7230	9520	960
1-20-D	Clay	5.60	--	4485	11779	82608	289300	0
1-21-D	Lignite	5.73	54.8	3510	6974	50912	70280	0
1-22-D	Claystone	5.81	--	4900	10756	87974	291200	0
1-23-D	Mudstone	6.41	--	7708	9368	64715	285100	0

*Height from base of major coal seam.
**Determined by x-ray fluorescence.
~Moisture free basis.

and Minor Elements, Parts Per Million (Dry Basis)

S**	Cl	K	Ca	Sc	Ti	V	Cr	Mn	Fe	Co	Ni
370	411	26788	9605	15	4052	137	99	144	26761	5	59
14370	132	2153	6247	6	724	219	49	39	9377	2	34
16880	55	806	5984	2	246	25	6	49	6685	2	21
8910	44	1000	7658	1	119	4	2	69	2909	1	16
6990	59	615	6888	1	323	22	5	47	2326	1	12
6090	55	1000	6820	1	124	4	1	68	1745	0	13
9480	67	1000	6994	1	298	16	5	76	2182	1	16
7790	36	1000	4689	0	78	1	1	24	6128	1	13
10820	74	1104	6330	3	357	47	10	61	7323	4	28
750	407	1000	8691	11	2578	101	68	100	17133	3	33
11030	51	1000	7935	7	191	12	5	56	4070	1	33
3000	336	25718	10564	13	3119	138	80	257	29506	11	47
1130	407	1000	16988	11	3393	116	72	287	24822	11	36
1130	508	1000	7838	15	3540	170	94	175	31466	15	55
0	116	5000	5824	29	4320	138	21	128	233070	1	190
21120	113	1000	6754	15	419	185	34	66	28510	10	46
1130	428	30645	7483	14	4103	165	93	187	30072	16	49
11270	43	233	6391	4	125	6	4	79	3552	1	25
13960	63	1000	6550	3	145	24	10	108	3724	4	24
370	471	1000	12364	14	3869	134	75	424	30382	14	67
9090	314	21576	7039	11	2133	158	92	242	28376	13	64
1120	499	1000	8418	14	3429	151	106	387	30294	19	96
750	395	19274	18107	13	3081	106	119	436	30374	15	50

Continued on next page

Table IV.

ID		Cu*	Zn	Ge	As	Se	Br	Rb**	Sr**	Y**	Zr**	Ru
1-1-D	Underclay	68	42	--	4	2	1	168	255	28	93	160
1-2-D	Lignite	259	25	11	16	2	1	11	250	7	146	16
1-3-D	Lignite	147	14	0	20	1	3	9	346	5	21	5
1-4-D	Lignite	42	10	0	6	1	2	5	353	5	10	3
1-5-D	Lignite	77	6	0	6	1	2	0	207	5	14	3
1-6-D	Lignite	37	8	0	5	0	2	5	400	5	11	3
1-7-D	Lignite	32	12	0	8	1	3	5	419	5	24	3
1-8-D	Lignite	17	6	0	31	0	1	0	189	5	10	3
1-9-D	Lignite	74	19	0	15	1	1	0	201	5	109	5
1-10-D	Clay parting	67	34	--	7	2	1	117	217	20	60	119
1-11-D	Lignite	83	25	0	21	1	4	5	276	7	72	6
1-12-D	Clay	79	44	--	8	2	1	155	193	29	120	147
1-13-D	Mudstone	70	36	--	3	2	3	91	157	26	121	107
1-14-D	Claystone	98	54	--	3	2	1	127	149	25	86	144
1-15-D	Concretion Zone	12	108	--	2	9	3	25	38	23	17	46
1-16-D	Lignite	79	40	57	28	2	2	10	260	5	381	9
1-17-D	Clay	110	45	--	8	2	1	129	148	25	74	138
1-18-D	Lignite	83	20	--	17	1	3	5	229	6	17	4
1-19-D	Lignite	180	18	--	27	1	3	0	194	5	24	4
1-20-D	Clay	78	47	--	2	2	1	133	156	28	94	122
1-21-D	Lignite	159	37	--	44	4	3	33	99	10	247	66
1-22-D	Claystone	107	46	--	7	2	2	134	153	25	83	131
1-23-D	Mudstone	83	38	--	11	1	2	81	164	22	108	90

*Height from base of major coal seam.
**Determined by x-ray fluorescence.

-- Continued

Ag	Cd	Sb	Cs	Ba	La	Ce	Sm	Eu	Yb	Th	U
2.48	3	2.11	5.05	1132	27	53	3.07	0.78	1.27	9.75	4.24
1.42	5	19.10	4.40	370	9	18	1.86	0.51	1.19	4.01	4.51
0.83	9	4.82	1.34	185	3	9	1.61	0.16	0.27	1.16	6.12
0.60	5	0.62	0.14	64	3	5	0.58	0.08	0.11	0.38	0.49
0.71	5	2.27	0.46	39	3	4	0.38	0.05	0.22	0.92	0.65
0.49	5	0.19	0.39	45	2	3	0.25	0.05	0.08	0.34	0.41
0.67	4	1.31	0.29	66	4	6	0.62	0.10	0.12	1.43	1.67
0.46	1	0.7	0.21	496	1	4	0.24	0.04	0.18	0.51	0.35
1.12	9	5.23	1.17	114	4	8	1.62	0.36	0.52	2.39	5.29
2.05	5	0.85	5.77	745	26	50	2.25	0.59	1.34	11.02	3.61
1.44	7	3.77	0.37	246	5	9	1.56	0.26	1.02	2.51	4.07
2.57	5	2.45	4.05	661	35	58	3.99	0.95	1.65	10.15	4.09
2.19	5	1.17	2.79	673	31	60	3.49	1.21	1.92	11.07	2.03
3.11	5	1.73	4.75	781	32	63	3.53	1.12	1.80	10.99	2.66
7.14	20	0.31	0.44	595	23	6	3.16	0.87	2.99	2.81	2.50
2.33	18	15.11	5.60	179	5	1	2.61	0.48	1.19	1.61	8.56
2.00	5	2.03	0.73	761	32	56	3.15	0.96	1.68	10.08	3.10
1.09	0	4.54	0.86	161	3	6	1.02	0.2	0.65	0.72	0.70
1.00	4	3.79	0.81	95	2	4	1.15	0.22	0.41	1.73	3.10
2.74	5	1.33	3.58	640	29	57	3.22	1.01	1.76	10.22	2.12
2.21	8	7.77	3.75	559	19	36	2.52	0.90	1.67	7.69	5.40
2.75	2	1.67	3.84	911	26	49	3.01	1.02	1.37	9.37	2.36
2.30	5	2.04	2.24	691	29	55	3.3	1.14	1.83	9.18	1.75

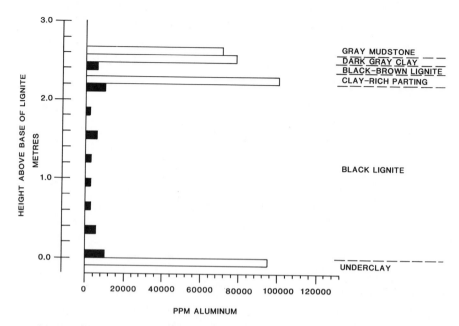

Figure 2. Aluminum distribution in the lower part of the lithologic sequence including underclay, the Kinneman Creek lignite,
and directly overlying clay and mudstone.

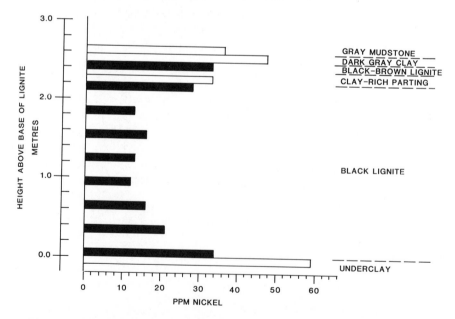

Figure 3. Nickel distribution in the lower part of the lithologic sequence including underclay, the Kinneman Creek lignite,
and directly overlying clay and mudstone.

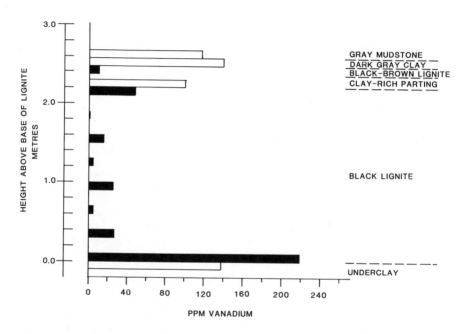

Figure 4. Vanadium distribution in the lower part of the lithologic sequence including underclay, the Kinneman Creek lignite, and directly overlying clay and mudstone.

Creek Bed include Na, Mg, Ca, Sr, and Mn. Figures 5 and 6 illustrate the distribution of Sr and Ca in the Kinneman Creek Bed and adjacent sediments. Although some variation in the concentrations is noted from sample to sample, the criteria for being considered to have an even distribution were that the concentrations were colinear when plotted on probability paper and that none of the points could be rejected as outliers by statistical tests (17).

The elements with irregular distribution include P, Br, and Cd.

Interpretation of Elemental Distribution Within the Coal Seam.   The observed patterns of element distribution in the lignite seam may be explained by both changes in depositional conditions during accumulation and changes during diagenesis and post-diagenetic processes. For example, depositional factors involving addition of greater fractions of detrital clay and silt at the beginning and end of peat deposition would increase Si and Al in the margins of the seam. Potassium is concentrated in the base of the major seam and may be associated with detrital minerals characteristic of the underclay and immediately overlying lignite.

The possible associations of trace and minor elements with clays are summarized by Finkelman (14). These include Al, Sc, Ti, Cl, Cr, La, Mg, and Fe. Iron, Co, Ni, Zn, As, and Ag all have chalcophile tendencies and may associate with pyrite or other sulfides. Uranium may be largely associated with the organic part of the coal but can also be associated with a diverse suite of uranium minerals (14). Other elements, such as Ru, Ce, and Sm are concentrated in the margins of the seam suggesting an association with detrital minerals.

Other possible depositional factors that can effect the concentration or mode of occurrence of an element include change of Eh or pH, influx of ash, or change of botanical factors at the beginning and end of peat deposition. Post-depositional factors related to the flow of meteoric water through the lignite laterally might selectively concentrate elements in the margins or its vertical flow might concentrate elements at either the upper or lower margin, depending on flow direction and on the permeability of the adjoining sediments. Other post-depositional factors might be related to the changing geochemistry of the lignite-forming environment, such as Eh and pH changes, degradation of plant material, and geological factors such as depth of burial, temperature, compaction, or changes in groundwater chemistry.

Elements which showed a clear tendency to concentrate toward the base of the lignite seam include V, Sb, and Cs. Vanadium has strong organic tendencies (14). Antimony has strong chalcophile tendencies so it is possible that Sb is associated with the sulfides.

The relationship between the concentration of an element in "whole" coal and the ash content can be used as a guide to the affinity of that element for, or incorporation in, the mineral matter or the carbonaceous material. If the concentration of an element increases with increasing ash content that element is presumed to be associated with the inorganic species that form ash, or in other words may be said to have an inorganic affinity. If the concentration shows no correlation with ash content, that element would be said to have an organic affinity. Linear least squares correlation coefficients were calculated for the concentrations of 39 elements

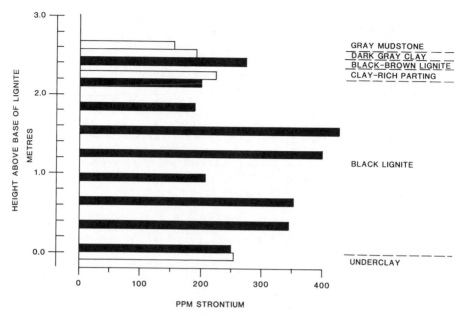

Figure 5.   Strontium distribution in the lower part of the lithologic sequence including underclay, the Kinneman Creek lignite,
and directly overlying clay and mudstone.

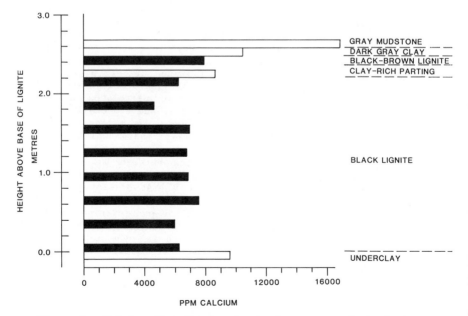

Figure 6.   Calcium distribution in the lower part of the lithologic sequence including underclay, the Kinneman Creek lignite,
and directly overlying clay and mudstone.

(deleting Ge) listed in Table III versus the ash content (moisture-free basis) for 12 of the 13 lignite samples. Sample 1-21-D was not included because it had 54.8% mf ash, which is clearly well outside the range of ash content in the other samples. Seven elements - Na, Ca, Mn, Br, Sr, Y, and Ba - had correlation coefficients below 0.200 and thus show organic affinity in this suite of samples. An additional seven elements - Mg, K, Cu, As, Rb, Ce, and Eu - had correlation coefficients ranging from 0.201 to 0.600 and may be associated with both the carbonaceous and mineral portions of the coal. The remaining 24 elements show inorganic affinity.

A more restricted study of 28 elements for the main lignite seam (samples 1-2-D through 1-9-D) showed generally similar results. Calcium, Mn, Sr, As, and Ba had correlation coefficients of <0.200. Elements with correlation coefficients >0.600 were Al, Si, K, Sc, Ti, V, Cr, Fe, Co, Ni, Se, Sb, Cs, La, Ce, Sm, Eu, Th, and U. Elements having mixed organic-inorganic association ($0.200 \leq r \leq 0.600$) were Na, Mg, Cu, and Rb.

Reasonable qualitative agreement exists among the elemental variations observed here, the ionic potential of the elements, and the "chemical fractionation" behavior. Chemical fractionation determines the amount of each element present as ion-exchangeable cations, as acid-soluble minerals or coordination complexes, and as insoluble minerals. Our application of this procedure to other lignites has been discussed previously (16). In general, those elements which would be predicted to form insoluble hydrolysates on the basis of their ionic potential (that is, $3 < Z/r < 12$, where Z is the ionic charge and r the radius) are found concentrated near the margins of the seam. During chemical fractionation these elements are mainly insoluble in 1M HCl. Examples of elements in this category are Al, Ti, Cr, and Ni. On the other hand, elements of $Z/r < 3$ would likely exist as hydrated cations; these elements generally show even distribution or weak concentration in the seam and are removed by ion-exchange with ammonium acetate. Examples are Na, Mg, and Ca. Most of the elements for which $Z/r < 3$ also show organic affinity. The even distribution suggests that during the geological history of the seam these elements have somehow washed or percolated through the seam, undergoing ion exchange with the coal and eventually coming to an apparent equilibrium throughout the seam. In a sense the lignite seam could be thought of as an ion-exchange medium which has become "loaded" by the solution passing through it.

A relationship can be demonstrated between the distribution observed in the seam and the geochemical properties of the elements. The relationship is summarized in Table V. At present the information in Table V is based only on the study reported here, and therefore must be considered to be a qualitative presentation of rules of thumb. Current work in our laboratory is focused on several additional lithologic sequences. As more data is accumulated we expect to be able to extend Table V and incorporate quantitative relationships.

Table V. Qualitative Relationships Between Geochemical Properties and Distribution in Seam

Form Predicted From Ionic Potential	Inorganic/ Organic Affinity	Chemical Fractionation Behavior	Distribution in Seams	Examples
Cation	Organic	Principally ion-exchangeable	Even	Na,Ca
Cation	Inorganic	Occurs in all three fractions	Concentrates at bottom	K
Insoluble hydrolysate	Inorganic	Acid-soluble and residual forms only	Concentrates at both margins	Al, Si

Variation Within Other Components of the Lithologic Sequence.   The distribution of major and minor elements in the lithologic sequence outside of the main coal seam is generally consistent with the mineral content of the sediments. Clay- and silt-rich sediments are enriched in Al, Si, and possible Na, Mg, K, and/or Ca. The sideritic concretionary zone at 3.04 m (1-15-D) is enriched in Fe and other elements including Sc, Ni, Zn, Se, Ag, Cd, and Yb (Table IV). Elements depleted in the concretion zones relative to adjacent overburden include Na, Mg, Cl, Ca, Cr, Mn, Co, Cu, Rb, Sr, Zr, Ru, Cs, Ce, and Th. The thin coal seam at 4.75 m (1-16-D) has relatively high concentrations of V, Ge, As, Cd, Sb, and U (Table IV). Specific beds, including the underclay at -0.05 m (1-1-D), are markedly enriched in K, possibly related to a high mica or illite content.

Many of the minor metallic elements have concentrations lying within 20% of the average values for shales. This comparison is illustrated by the data in Table VI, which compares the average concentrations in the underclay and overburden samples, after statistical rejection of outliers by the Dixon method (17), with tabulated values from the literature (18). The most notable exception is manganese, for which the average value in shales is 850 ppm but the observed average in the overburden and underclay was 269 ppm.

The variation of potassium in the overburden and underclay is extreme, ranging from 1000 to 30,000 ppm. The reasons for this variability are not clearly understood, but may be due to the distribution minerals in the overburden and underclay, some of which are potassium-containing clays.

Table VI.   Comparison of Average Minor Element Concentrations (ppm)
for Overburden/Underclay Samples with Average Concentrations for
Shales

Element	Overburden/Underclay	Shales	Element	Overburden/Underclay	Shales
Ba	760	600	Sc	15	15
Ce	56	70	Th	10	12
Co	13	20	Ti	3650	4600
Cr	92	100	U	2.8	3.5
Mn	269	850	V	139	130
Ni	72	80	Zn	51	90

Acknowledgments

The authors would like to thank Diane K. Rindt for the x-ray diffraction work and Arthur L. Severson for x-ray fluorescence analysis.
Mr. Roaldson acknowledges support from Associated Western Universities, Inc., as part of their summer undergraduate research participation program.

This report was prepared as an account of work sponsored by the
United States Government.   Neither the United States nor any agency
thereof, nor any of their employees, makes any warranty, express or
implied, or assumes any legal liability or responsibility for the
accuracy, completeness, or usefulness or any information, apparatus,
product, or process disclosed, or represents that its use would not
infringe privately owned rights.   Reference herein to any specific
commercial product, process, or service by trade name, mark, manufac-
turer, or otherwise, does not necessarily constitute or imply its
endorsement, recommendation, or favoring by the United States Government or any agency thereof.   The views and opinions of authors
expressed herein do not necessarily state or reflect those of the
United States Government or any agency thereof.

Literature Cited

1.   "Analysis of the Northern Great Plains Province Lignites and
     Their Ash: A Study of Variability,"   U.S. Bureau of Mines
     Report RI 7158, 1968.
2.   "Low-rank Coal Study, National Need for Resource Development-
     Resource Characterization," U.S. Department of Energy Report
     DOE/FC/10066-TI (Vol. 2), 1980.
3.   "Geology of the Upper Part of the Fort Union Group (Paleocene),
     Williston Basin, with Reference to Uranium,"   North Dakota Geol.
     Survey Report of Investigation No. 58, 1976.

4.   "Geology and Geohydrology of the Knife River Basin and Ajacent Areas of West-Central North Dakota," North Dakota Geol. Survey Report of Investigation No. 64, 1979.

5.   "Geologic Sampling of Lignite Mines in Mercer and Oliver Counties, North Dakota," U.S. Department of Energy Report DOE/FC/FE-1, (in preparation, 1983).

6.   Logan, K.J.   M.S. Thesis, University of North Dakota, Grand Forks, 1981.

7.   Karner, F.R.; Beckering, W.; Rindt, D.K.; and Schobert, H.H.; Geol. Soc. Amer. Abstracts with Programs 1979, 11, 454.

8.   Karner, F.R.; Winbourn, G.; White, S.F.; Gatheridge, A.K.; N.D. Acad. Sci. Proceedings 1979, 33, 70.

9.   Schobert, H.H.; Benson, S.A.; Jones, M.L.; Karner, F.R.; Proc. 8th International Conference on Coal Sciehce, 1981, p. 10-15.

10. Karner, F.R.; Benson, S.A.; Schobert, H.H.; Roaldson, R.G.; Critchlow, K.; Geol. Soc. Amer. Abstracts with Programs 1982, 14, 525.

11. Weaver, J.N.; in "Analytical Methods for Coal and Coal Products"; Karr, C., Jr. (Ed.); Academic Press, Inc.; New York, 1978; Vol. I, p. 377-401.

12. Benson, S.A.; N.D. Acad. of Sci. Proceedings 1980, 34, 10.

13. "A Geochemical Study of the Inorganic Constituents in some Low-Rank Coals," U.S. DOE Report FE-2494-TR-1, 1979.

14. Finkelman, R.B.   Ph.D. Thesis, University of Maryland, College Park, 1981.

15. "Distribution of Inorganics in Selected Low-Rank Coals," Grand Forks Energy Technology Center Internal Report GFETC IR-5, 1982.

16. Benson, S.A.; Holm, P.L.; ACS Div. of Fuel Chem. Preprints 1983, 28(2), 234-9.

17. Lipson, C.; Sheth, N.J.   "Statistical Design and Analysis of Engineering Experiments"; McGraw-Hill Book Co.: New York, 1973; p. 91.

18. Krauskopf, K.B.   "Introduction to Geochemistry"; McGraw-Hill Book Co.: New York, 1979; p. 482.

RECEIVED April 1, 1984

Measurement and Prediction of Low-Rank Coal Slag Viscosity

ROBERT C. STREETER[1], ERLE K. DIEHL[1], and HAROLD H. SCHOBERT[2]

[1]Bituminous Coal Research, Inc., National Laboratory, Monroeville, PA 15146
[2]Energy Research Center, University of North Dakota, Grand Forks, ND 58202

Viscosities of 17 western U.S. lignite and subbituminous coal slags were measured over the temperature range
of 2100° to 2700°F (1150° to 1480°C) using a rotating-
bob viscometer. Efforts to correlate the viscosity/
temperature data with coal ash or slag composition,
using correlations developed in England for higher-rank
bituminous coals, were generally unsatisfactory for the
low-rank coal slags, presumably due in part to the
higher amounts of alkali and alkaline earth metals normally present in western U.S. coals. Two more recent
correlations, developed in France for metallurgical
slags, were also treated. One of these, reported by G.
Urbain et al., is based on the SiO_2-Al_2O_3-CaO ternary
phase diagram. When the low-rank-coal slags were
subdivided into three groups, based on a parameter of
the Urbain equation that is roughly proportional to
silica content of the slag, modified forms of the
Urbain equation gave acceptable correlations (within
30%) for nearly 50% and marginally acceptable correlations (within 60%) for 70% of the viscosity tests.

Since 1980, BCR National Laboratory (BCRNL) has been conducting
measurements on the viscosity of western U.S. low-rank coal slags
under a contract within the U.S. Department of Energy. This work has
been motivated by the realization that 1) very few data exist in the
literature on the viscosity of low-rank-coal slags; 2) published correlations relating slag viscosity to coal ash composition were derived
from work with bituminous-coal slags, and attempts to apply them to
data from low-rank coal slags generally have been unsuccessful; and,
3) from a practical standpoint, data on slag rheology are of considerable interest in support of the operation of slagging coal gasifiers.
Although the data obtained in these studies have been valuable
in interpreting the slagging phenomena observed in gasifier tests, a
more fundamental objective of the current work is to develop correlations that can be used to predict a priori the viscosity behavior of
low-rank-coal slags from a knowledge of the ash or slag composition.

0097-6156/84/0264-0195$06.00/0
© 1984 American Chemical Society

Experimental

The slag viscosity apparatus was assembled by Theta Industries, Inc. It includes a Lindberg furnace rated for operation at temperatures up to 2732°F (1500°C). The rotating bob viscometer is a Haake RV-2 Rotovisco unit with a DMK 50/500 dual measuring head. The torque on the viscometer bob twists a spiral spring inside the measuring head; the angular displacement of the spring is proportional to shear stress, and is converted to an electrical signal that is plotted on an X-Y recorder as a function of the rotational speed of the bob (shear rate).

To simulate conditions existing in a slagging coal gasifier, all tests were carried out in a reducing atmosphere of 20% H_2 and 80% N_2. The gas mixture is injected into the furnace at a flow rate of 500 cc/ min through an alumina tube that extends to within 1 inch of the top of the sample crucible.

In all of the tests discussed herein, the bob was fabricated from half-inch molybdenum bar stock; the bob is approximately 1 inch long with a 30° angle taper machined on both ends. The bob stem is covered by a tubular alumina sleeve to minimize erosion during a test.

In the earlier tests, the slag was melted in crucibles of vitre- ous carbon. It soon become evident that iron oxides in the slag were being reduced by reaction with the carbon crucible, as small pools of molten iron invariably settled from the slag during the carbon- crucible tests. Consequently, the carbon crucibles were eventually replaced with high-purity alumina crucibles.

Since the alumina crucibles were put into use, metallic iron has only rarely been observed in the slag. On the other hand, varying degrees of attack on the alumina crucible by the slag have been observed. Generally, dissolution of Al_2O_3 by the slag has been slight, but in a few cases noticeable thinning of the crucible walls has occurred.

Since measured viscosities will vary depending on the dimensions of and the materials employed for the sample crucible and the rota- ting bob, measured values are related to absolute viscosities by means of an instrument factor. In these studies, the instrument factor was determined by tests with National Bureau of Standards glass viscosity standards whose viscosities are precisely defined and similar to those of the slags over the temperature range of interest.

In preparation for a slag viscosity test, the coal is pulverized to minus 60 mesh and ashed for 3 hours at 1850°F (1010°C) in a large muffle furnace. This higher-than-normal ashing temperature was chosen after preliminary work showed that sufficient volatiles remain- ed in the ash prepared at the standard ASTM temperature of 1382°F (750°C) to cause severe frothing and foaming during the ash melting process. The coal ash is sieved again at 60 mesh to remove occasional adventitious particles of firebrick, then compressed into pellets 5/8 inch in diameter. The ash pellets are dropped, one or two at a time, into the previously heated sample crucible through a long quartz tube; a typical coal-ash charge in the alumina crucibles is about 60 to 70 grams.

Viscosity measurements are normally commenced at the highest temperature at which an on-scale reading can be obtained with a bob speed of 64 rpm. The temperature is then decreased slowly (under microprocessor control) and viscosity measurements are taken at approximately half-hour intervals. This "cooling cycle" is continued until the upper limit of the measuring head is approached (typically about 3000 poises), then the slag is gradually reheated to observe changes in viscosity during a "heating cycle."

Following each test, samples of the solidified slag and high-temperature ash are analyzed by X-ray fluorescence and X-ray diffraction techniques. A Kevex Model 0700 energy dispersive X-ray spectrometer is employed for X-ray fluorescence analysis. X-ray diffraction measurements are conducted using a Philips 3600 automated X-ray diffractometer.

The selection of low-rank coals for these studies was based to some extent on those which had actually been tested or were candidate feedstocks for a slagging lignite gasifier; in addition, an effort was made to choose samples that would represent a wide range of ash compositions and diverse geographical locations. The coals selected for these viscosity studies are listed in Table I. In view of the considerable within-seam or within-mine variability of low-rank coals, it should be understood that these coal samples are not necessarily representative of the total production of a particular mine.

Table I. Samples Selected for Low-Rank Coal Viscosity Studies

Sample	Rank	Location (State)
Indian Head	Lignite	North Dakota
Gascoyne	Lignite	North Dakoa
Baukol-Noonan	Lignite	North Dakota
Beulah	Lignite	North Dakota
Colstrip	Subbituminous	Montana
Decker	Subbituminous	Montana
Sarpy Creek	Subbituminous	Montana
Naughton	Subbituminous	Montana
Big Horn	Subbituminous	Wyoming
Kemmerer	Subbituminous	Wyoming
Black Butte	Subbituminous	Wyoming
Emery	Bituminous (Low-Rank)	Colorado
Rockdale	Lignite	Texas
Martin Lake	Lignite	Texas
Big Brown	Lignite	Texas
Atlantic Richfield	Lignite	Texas
Burns & McDonnel	Lignite	Alabama

Results and Discussion

Viscosity can be defined as the resistance to flow offered by a fluid. According to Newton's law of viscosity, the applied shear

stress is proportional to the shear rate, the constant of proportionality being the absolute viscosity. The most commonly encountered
regime of viscous flow is termed Newtonian flow, in which a linear
relation exists between shear stress and shear rate. For many simple
liquids, the dependence of viscosity (η) on temperature (T) is expo-
nential, such that a plot of log η versus 1/T yields a straight line.
Three regimes of non-Newtonian behavior are less commonly encountered:
1.   Pseudoplastic - viscosity decreases with increasing shear rates.
2.   Plastic - viscosity decreases with increasing shear rates following the appearance of an initial yield stress.
3.   Thixotropic - viscosity decreases with increasing shear rates
and with the duration of the applied shear stress.
      Plots of log η versus T (°F) for three representative low-rank
coal slags are shown in Figure 1. Curve A for the Black Butte sample
illustrates the behavior of a glassy, siliceous-type slag that exhibits Newtonian properties over a wide temperature range. In addition,
the viscosity behavior is reversible in that viscosities measured
during reheating of the slag (solid circles) are essentially the same
as those determined during the cooling cycle. The type of behavior
represented by Curves B and C is more commonly observed for low-rank
coal slags; log η is a linear function of temperature down to some
point, designated as the temperature of critical viscosity, T_{cv}, (as
shown for Curve B). Below T_{cv}, the viscosity can increase quite
abruptly (Curve B), or the increase may be somewhat more gradual
(Curve C). Moreover, a "hysteresis" effect appears in the viscosity/
temperature curve as the slag is reheated. This phenomenon is sup-
posedly related to the slow redissolution of crystals, formed on
cooling the slag below T_{cv} ([1]). In theory, the viscosity of the slag
on reheating should eventually return to the linear portion of the
curve representing the fully liquid condition. However, as shown by
Figure 1, this is not necessarily achieved experimentally, and the
extent of the deviation is believed to be due to small compositional
changes occurring in the melt during the viscosity determination (to
be discussed later). Curve C in Figure 1 includes data points for
two separate viscosity tests with Baukol-Noonan, approximately 6
months apart. As shown, the repeatability for the cooling curve is
quite good; on the other hand, there is a marked difference in the
two curves obtained on reheating the slags, and subsequent analytical
data showed subtle differences in the compositions of the slags from
the two tests.
      The straight-line portion of the viscosity/temperature curve is
traditionally designated as the "Newtonian" region of slag behavior,
while the area below T_{cv}, where solid species are crystallizing from
the melt, is designated as the "plastic" region. Strictly speaking,
a slag can retain Newtonian properties below T_{cv}, and the transition
to a non-Newtonian state can be detected only by a change in the
shape the shear-stress vs. shear-rate curve. As a matter of fact,
most of the slags included in this study actually did show pseudoplastic or thixotropic behavior at the higher viscosities.
      Of the 17 coals listed in Table I, 13 were tested in carbon
crucibles and the remaining 4 were tested using alumina crucibles.
In addition, the carbon-crucible test with Baukol-Noonan was repeated, and six of the coals from the carbon-crucible tests (Big Horn,
Decker, Emery, Colstrip, Rockdale, and Burns & McDonnell) were also

APPARENT VISCOSITY (η), POISES

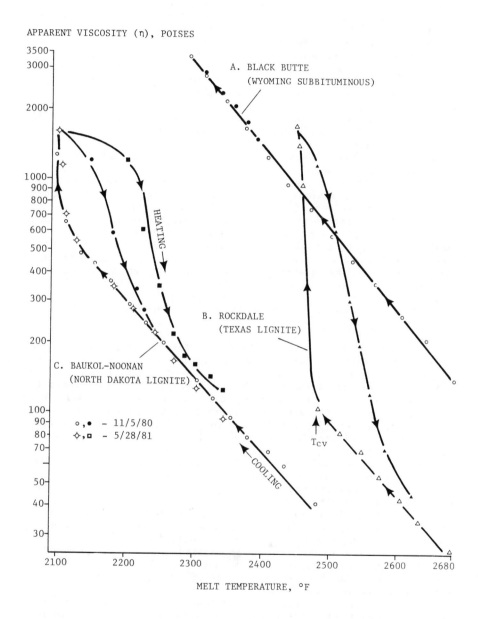

Figure 1. Representative slag viscosity curves for three
western U.S. low-rank coals.

selected for alumina-crucible tests. Thus, a total of 24 slag-
viscosity tests provided the data discussed in this paper.

The compositions of the slags from these viscosity tests are
listed in Table II. (The compositional data are normalized to 100%;
the median closure on raw analytical data was 94.2%.) Also included
are several other parameters that traditionally have been employed to
characterize coal-ash slags. It should be emphasized that, unless
otherwise noted, all calculations discussed herein were based on slag
composition data (as opposed to ash composition data). This was done
primarily for two reasons:

1. As previously noted, reactions with the sample crucible
tended to result in depletion of iron oxide species (carbon-crucible
tests) or enrichment in Al_2O_3 (alumina-crucible tests), so that in
some cases the composition of the slags was significantly different
from that of the original coal ash.

2. Results of an independent study, comparing high-temperature
ash and slag compositions for 18 of the 24 viscosity tests, showed
that from 80% to 100% of the SO_3 and usually all of the P_2O_5 (if
present in the ash) were volatilized during melting of the ash.
Admittedly, P_2O_5 is a minor constituent, and losses of SO_3 could be
compensated for by normalizing the analytical data to a sulfur-free
basis. On the other hand, in 11 out of 15 instances (73%) where Na_2O
was present in the high-temperature ash, it was volatilized in
amounts ranging from 10% to 50% of the amount available in the ash.
Since the Na_2O content of some ashes ranged up to 10%, it is probable
that volatilization of this constituent would have an effect on the
slag viscosity.

Slag Viscosity/Composition Correlations. Several attempts have been
made in the past to define the linear portion of the viscosity/
temperature curve based on the composition of the coal ash. In the
mid-1960's, workers at the British Coal Utilization Research Association (BCURA) developed two such correlations based on work with
British (bituminous) coals, now generally referred to as the Watt-
Fereday (4) and S^2 (3) correlations (6). Unfortunately, attempts to
apply this correlation to low-rank coal slags, using either ash or
slag composition data, have been generally unsuccessful.

Among the possible reasons why these predictive equations fail
for low-rank coals are the following:

1. Ash constituents may fall outside the concentration range
of those in most bituminous coals.

2. The BCURA predictive equations are based on ash analyses.
As discussed earlier, significant losses of certain elements may
occur as the ash from low-rank coals is heated to the melting point.

3. Sulfur retention is more prevalent in low-rank coal ashes,
due to the generally higher calcium contents. Some of the slags from
these viscosity studies retained up to 2% SO_3 (Table II), even after
being heated above 2600°F (1427°C).

4. In lignites, some 30% to 50% of the ash-forming constituents can consist of cations attached to the organic matter in ion-
exchangeable form on carboxyl groups or as chelate complexes, rather
than being present as distinct mineralogical species (5).

The Watt-Fereday equation is approximately a linear equation of
the form

$$\log \eta = m \cdot f(T) + c \tag{1}$$

where m is normally positive and c is negative. Revised values for the slope (m') and intercept (c') can be calculated based on the experimental data for the low-rank coals to obtain a good fit to the Watt-Fereday equation. However, for 20 of 23 tests (87%), the value of m' was larger than that predicted by the conventional Watt-Fereday equation; similarly, absolute values of c' were also larger for 16 of 23 tests (70%). Since the intercept is negative, these two effects should tend to offset each other, resulting in little change in the value of log η. In reality, the net result tended to be a larger value for log η, such that observed viscosities were higher than predicted viscosities in 18 of the 23 tests (78%).

Other combinations of variables, including those listed in Table II, were employed in efforts to derive empirical correlations with slag viscosity, but again without much success. In addition, the BCURA S^2 correlation (3) and a modified form of the Watt-Fereday correlation developed at the National Bureau of Standards (6) were tested. Of the 23 viscosity tests for which slag composition data are available, the NBS correlation gave a good fit to the experimental data in only two cases, while the S^2 correlation applied in only one case. Generally, the NBS and S^2 calculations gave higher viscosities than those predicted from the conventional Watt-Fereday treatment; the S^2 values were larger for 22 (96%) of the 23 tests, and the NBS values were larger for 19 (83%) of the tests. On the other hand, all three correlations had a tendency to underestimate the slag viscosity, as observed viscosities were greater than those estimated by any of the three predictive methods in 14 (61%) of the tests. Consequently, we conclude that the BCURA methods (or the NBS modification thereof) are generally unsatisfactory for estimating the viscosities of low-rank coal slags.

More recently, two correlations have appeared in the literature that were developed in France in studies with metallurgical (steel making) slags: The IRSID correlation, published by Riboud et al., (7) and the Urbain correlation (8). These two correlations are conceptually more appealing because:

o Unlike the BCURA correlations, which are based on the Arrhenius form of the viscosity equation

$$\eta = A \exp(E/RT) \tag{2}$$

the IRSID and Urbain correlations are based on the Frenkel relation

$$\eta = AT \exp(B/T) \tag{3}$$

which, according to the authors, is the preferred form.

o All slag constituents are included, rather than just the five major oxides.

o Slag constituents are expressed as mole fractions instead of weight percentages.

o The Urbain correlation is based on the SiO_2-Al_2O_3-CaO pseudoternary phase diagram, and these 3 constituents usually predominate in low-rank-coal slags.

Table II. Analysis of Low-Rank-Coal Slags Produced in Slag Viscosity Tests

Carbon Crucible Tests: Sample	Slag Composition, Weight Percent										Base/ Acid Ratio*	Silica Ratio+	SiO_2/ Al_2O_3 Ratio	Lignite Factor+
	SiO_2	Al_2O_3	Fe_2O_3	TiO_2	P_2O_5	CaO	MgO	Na_2O	K_2O	SO_3				
Gascoyne	42.4	14.8	3.1	1.4	0.0	25.7	8.2	3.7	0.1	0.2	0.70	53.4	2.86	10.9
Baukol-Noonan 1	41.4	19.2	1.0	2.3	0.0	21.8	4.3	7.7	0.6	1.2	0.56	60.4	2.16	26.1
Baukol-Noonan 2	39.3	19.7	4.5	1.7	0.0	18.3	4.4	9.9	0.6	1.1	0.62	59.1	1.99	5.0
Colstrip	45.0	23.9	2.1	1.8	0.0	20.6	5.9	0.1	0.1	0.4	0.41	61.1	1.88	12.6
Decker	38.4	20.0	6.5	2.1	0.0	18.7	4.2	8.6	0.3	0.9	0.63	56.6	1.92	3.5
Sarpy Creek	44.2	27.4	1.9	1.1	0.0	18.3	4.0	1.2	1.2	0.3	0.37	64.6	1.61	11.7
Naughton	63.5	17.8	5.5	0.8	0.2	6.7	3.6	0.0	1.6	0.1	0.21	80.1	3.57	1.9
Big Horn	27.1	28.6	11.4	1.5	0.7	16.3	6.9	5.0	0.5	1.8	0.70	43.9	0.95	2.0
Kemmerer	59.6	13.3	0.9	0.9	0.0	15.5	7.7	0.8	0.6	0.1	0.35	71.2	4.48	25.8
Black Butte	53.9	18.3	6.0	1.3	0.0	11.7	4.2	4.0	0.3	0.0	0.36	71.1	2.94	2.6
Emery	40.2	21.4	9.1	1.1	0.1	16.9	5.5	2.9	0.2	2.0	0.55	56.1	1.88	2.5
Rockdale	42.9	24.0	2.4	2.0	0.0	23.8	3.9	0.0	0.5	0.3	0.44	58.8	1.79	11.5
Atlantic Richfield	40.6	15.8	2.7	2.0	0.1	29.4	6.7	2.0	0.3	0.2	0.70	51.1	2.57	13.4
Burns & McDonnell	47.0	26.1	12.7	0.8	0.0	6.6	3.7	0.0	2.2	0.5	0.34	67.1	1.80	0.8

Alumina Crucible Tests:

Sample	Slag Composition, Weight Percent										Base/Acid Ratio*	Silica Ratio+	SiO₂/Al₂O₃ Ratio	Lignite Factor†
	SiO₂	Al₂O₃	Fe₂O₃	TiO₂	P₂O₅	CaO	MgO	Na₂O	K₂O	SO₃				
Indian Head	20.0	36.7	12.5	0.6	0.0	17.0	7.1	5.4	0.1	0.0	0.73	35.3	0.54	1.9
Beulah (high sodium)	24.2	33.6	14.0	0.9	0.0	14.4	4.3	8.1	0.2	0.3	0.70	42.5	0.72	1.3
Colstrip	38.0	28.1	11.2	1.5	0.0	17.1	3.9	0.0	0.1	0.0	0.48	54.1	1.35	1.9
Decker	36.1	23.0	8.8	2.1	0.0	18.2	3.8	7.5	0.2	0.0	0.63	54.0	1.57	2.5
Big Horn	30.2	24.7	13.6	1.8	0.8	18.1	5.5	4.5	0.5	0.1	0.74	44.8	1.22	1.7
Emery	44.2	24.7	7.4	1.4	0.0	15.1	4.1	2.6	0.2	0.0	0.42	62.4	1.79	2.6
Rockdale	38.5	27.5	8.9	1.7	0.0	20.0	2.7	0.0	0.4	0.0	0.47	54.9	1.40	2.6
Martin Lake	39.0	22.4	11.6	1.1	0.0	22.4	2.5	0.0	1.0	0.1	0.60	51.6	1.74	2.1
Big Brown	(analytical data for slag not yet available)													
Burns & McDonnell	45.7	23.1	18.9	0.8	0.0	6.0	2.8	0.0	2.0	0.2	0.43	62.3	1.98	0.5

* Base/Acid Ratio = $(Fe_2O_3 + CaO + MgO + Na_2O + K_2O)/(SiO_2 + Al_2O_3 + TiO_2)$ (2)

+ Silica Ratio = $100\ SiO_2/(SiO_2 + "Fe_2O_3" + CaO + MgO)$ where $"Fe_2O_3" = (Fe_2O_3 + 1.11\ FeO + 1.43\ Fe)$ (3)

† Lignite Factor = $(CaO + MgO)/Fe_2O_3$ (2). Theoretically, $\Sigma(CaO + MgO + Fe_2O_3)$ must be greater than 20% and the lignite factor must be greater than 1.0 for the ash (slag) to be considered a low-rank-coal type; the Burns & McDonnell sample is an obvious exception to this rule.

However, these correlations assume the oxidation state of iron in the slag is known. Since this information is not currently available for the low-rank-coal slags, and because all tests were in a reducing atmosphere, for computational purposes it was tentatively assumed that all iron was in the form of FeO. (In cases where metallic iron was formed as a separate phase, it was normally removed before the slag sample was subjected to X-ray analysis.)

On this basis, the IRSID correlation gave a reasonably good fit to the experimental data (within 20%) for 3 of the 23 viscosity tests. Unlike the BCURA correlations, which tended to give constituently low results, the error in the IRSID correlation tended to be random; predicted viscosities were too high in 11 cases and too low in 9 cases.

The Urbain correlation was acceptable for 6 tests, too high in 7 instances, and too low in 10 instances. Furthermore, even though the Urbain correlation was unsatisfactory for 17 tests, compared with the IRSID correlation, the predictive viscosities were closer to the actual values in 10 instances. Of all the correlations tested (including the BCURA correlations), the Urbain correlation was judged to give the closest agreement to the actual viscosities in 7 of the 23 tests; among the remaining 16 tests, the Urbain correlations was judged second-best for 9.

Since the Urbain method appeared to give a fair-to-good correlation for nearly two-thirds of the viscosity tests, efforts were directed toward modifying this procedure with the goal of optimizing the fit to the experimental data. The logarithmic form of the Urbain equation is

$$\ln \eta = \ln A + \ln T + 10^3 B/T \quad (T \text{ in } °K) \qquad (4)$$

where B is a parameter defined by the composition of the slag and A is a function of B. The equation can be "forced" to fit the experimental data by including a fourth term

$$\ln \eta = \ln A + \ln T + 10^3 B/T - \Delta \qquad (5)$$

where Δ is the difference (either positive or negative) between the actual and computed (by Equation 4) values of $\ln \eta$. Since $\ln \eta$ is a linear function of temperature above T_{cv}, Δ will also be a linear function of temperature:

$$\Delta = mT \ (°K) + b \qquad (6)$$

Thus, for each slag, values for m and b can be derived from the experimental viscosity data which uniquely define the linear portion of the viscosity/temperature curve by means of Equation 5. The process then becomes one of correlating the variables m and b with some particular property of the slag.

It was found that, to a first approximation, b could be correlated with m by means of the expression

$$b = -1.6870(10^3 m) + 0.2343 \qquad (7)$$

with a correlation coefficient R = -0.988 for all 23 viscosity tests. The process was then reduced to finding a correlation for the variable m. Unfortunately, however, no single correlation was found that satisfied the data from all the viscosity tests.

As a further refinement, the slags were subdivided into three groups based on the magnitude of the parameter B in the Urbain equation (Equation 4). The magnitude of this parameter is related to the location of the slag composition in the SiO_2-Al_2O_3-CaO ternary phase diagram, and it is proportional to the silica content of the slag. Thus, with certain borderline cases, the three subgroups correspond roughly to "high-silica," "intermediate-silica," and low-silica" slags.

The "high-silica" group included the five tests with Naughton, Burns & McDonnell, Black Butte, and Kemmerer (B>28), and a fair correlation (R = -0.971) was found for m of the form

$$10^3m = -1.7264F + 8.4404 \qquad (8)$$

where $F = SiO_2/(CaO + MgO + Na_2O + K_2O)$ and slag components are in mole fractions. With this correlation, calculated viscosities were within 10% (very good) of actual viscosities for one test, within 30% (acceptable) for two other tests, and within 60% (marginally acceptable) for the 4th of 5 tests. Similarly, for the "intermediate-silica" slags (Colstrip, Rockdale, Emery, Sarpy Creek, and Martin Lake; B = 24 to 28), an expression similar to Equation 8 was found. In this case, however, the correlating variable (F) appeared to be the product of the Urbain parameter B times the sum of the mole fractions of Al_2O_3 and FeO. For the 8 viscosity tests in this group, the modified correlation was very good for 2, acceptable for 2 others, and marginally acceptable for 2 more. For the "low-silica" slags (Gascoyne, Baukol-Noonan, Indian Head, Big Horn, Decker, Atlantic Richfield, and Beulah; B<24), the correlating variable chosen was $CaO/(CaO + MgO + Na_2O + K_2O)$, although the correlation was less satisfactory, being very good for 2 tests, acceptable for 2 others, and marginally acceptable for 2 more, out of 10 tests. The equations developed for the modified Urbain correlations are summarized below, and viscosities calculated using the various correlations are compared with measured viscosities in Table III. The two viscosity values shown in Table III for each sample, corresponding to specific melt temperatures, are sufficient to approximate the linear ("Newtonian") portion of the log η versus T curve.

High-Silica Slags:
$b = -1.7137(10^3m) + 0.0509$ (R = -0.990 for 5 of 5 data points)
$10^3m = -1.7264F + 8.4404$ (R = -0.971 for 5 of 5 data points)
where
$F = SiO_2/(CaO + MgO + Na_2O + K_2O)$, mole fractions.

Intermediate-Silica Slags:
$b = -2.0356(10^3m) + 1.1094$ (R = -0.998 for 7 of 8 data points)
$10^3m = -1.3101F' + 9.9279$ (R = -0.982 for 5 of 8 data points)
where
$F' = B(Equation 4) \times (Al_2O_3 + FeO)$

Table III. Observed and Calculated Viscosities for Low-Rank-Coal Slags

Carbon Crucible Tests:

Sample	Temp., °F	Observed Viscosity, poises	Viscosity (poises) Calculated from Predictive Correlations					
			Modified Urbain	Conventional Urbain	IRSID	Watt-Fereday	NBS-Modified Watt-Fereday	BCURA S^2
Gascoyne	2458	25	25	33	31	20	44	44
	2237	102	105	93	103	80	157	190
Baukol-Noonan 1	2480	41	40	48	56	54	101	87
	2133	482	524	296	500	822	1523	945
Baukol-Noonan 2	2343	93	273	83	91	119	197	176
	2128	548	2276	269	371	738	1220	833
Colstrip	2557	39	117	80	133	37	67	60
	2463	102	233	128	242	64	118	105
Decker	2346	50	158	78	94	84	134	129
	2147	289	911	228	346	414	651	536
Sarpy Creek	2697	56	45	60	83	27	43	44
	2491	220	177	169	318	86	144	139
Naughton	2731	979	611	667	264	262	180	365
	2553	3375	2018	2110	777	817	587	963
Big Horn	2407	93	130	44	67	10	22	23
	2272	325	477	85	172	22	48	57
Kemmerer	2637	88	118	121	98	116	149	151
	2413	589	753	415	341	514	649	566
Black Butte	2688	140	94	109	90	94	91	113
	2294	3375	2385	1059	1023	1494	1544	1227
Emery	2548	98	44	42	61	24	37	34
	2354	422	157	107	201	74	113	114
Rockdale	2681	26	41	38	55	16	29	23
	2481	105	165	97	185	44	82	70

Atlantic Richfield	2396 / 2183	27 / 172	12 / 28	42 / 120	47 / 163	20 / 82	46 / 165	50 / 222
Burns & McDonnell	2690 / 2535	94 / 272	105 / 304	119 / 272	160 / 448	54 / 129	47 / 118	64 / 151

Alumina Crucible Tests:

Indian Head	2594 / 2508	48 / 82	24 / 52	21 / 30	19 / 34	2 / 3	5 / 7	4 / 6
Beulah (high sodium)	2597 / 2380	38 / 340	31 / 249	24 / 65	24 / 103	5 / 13	8 / 25	7 / 24
Colstrip	2653 / 2520	18 / 34	16 / 30	43 / 80	70 / 166	12 / 23	18 / 35	15 / 32
Decker	2286 / 2146	69 / 237	214 / 690	124 / 273	180 / 488	91 / 283	145 / 451	143 / 399
Big Horn	2331 / 2177	90 / 289	101 / 344	67 / 150	91 / 268	17 / 48	31 / 81	41 / 124
Emery	2564 / 2441	53 / 114	66 / 145	96 / 183	138 / 309	51 / 107	68 / 146	68 / 142
Rockdale	2617 / 2553	25 / 36	24 / 34	50 / 68	87 / 131	16 / 21	24 / 33	20 / 29
Martin Lake	2679 / 2322	43 / 635	16 / 135	27 / 148	31 / 274	9 / 54	13 / 77	10 / 87
Burns & McDonnell	2698 / 2402	40 / 242	73 / 491	82 / 401	97 / 670	38 / 202	28 / 156	32 / 178

Low-Silica Slags:
  $b = -1.8244(10^3 m) + 0.9416$ (R = -0.999 for 6 of 10 data points)
  $10^3 m = -55.3649F'' + 37.9186$ (R = -0.970 for 7 of 10 data points)
where
  $F'' = CaO/(CaO + MgO + Na_2O + K_2O)$, mole fractions.

The nature of the "F" terms in the modified Urbain correlations is of interest. Urbain categorizes the constituents of silicate melts as "glass formers" (SiO_2, P_2O_5), "modifiers" (CaO, MgO, Na_2O, K_2O) and "amphoterics" (Al_2O_3, Fe_2O_3) which can act either as glass formers or modifiers. The correlating term for high-silica slags is the ratio of SiO_2 to modifiers, suggesting that, for these slags, the silica content of the slag is the dominating factor. On the other hand, for low-silica slags the correlating terms is the ratio of CaO to modifiers, which implies that the amount of CaO relative to total modifiers in the slag is a critical factor. For intermediate-silica slags, the correlating term involves the "amphoterics," suggesting that the role of these constituents becomes more important when the nature of the slag cannot be well defined by its silica content.

Summary and Future Work

Data from 23 low-rank coal slag viscosity determinations have been employed in attempts to correlate the linear portion of the log viscosity versus temperature curve with five published empirical correlations for silicate slags. Of these five correlations, one developed by Urbain for metallurgical slags appeared to give a reasonable fit to the experimental data for nearly two-thirds of the viscosity tests. When the slags were subdivided into three groups, based on a parameter of the Urbain equation that is roughly proportional to silica content, it was possible to derive modified forms of the Urbain equation that gave acceptable correlations (within 30%) for 11 of the viscosity tests. Of the remaining 12 tests, the modified correlations were marginally acceptable (within 60%) in 5 cases.
Efforts are still being made to refine these empirical correlations and, in particular, to understand the reasons why certain slags fail to fit the correlations. The ultimate objective is to interpret slag rheological behavior in terms of specific phases present in the slag, to the extent that such phases can be identified and quantified.

Acknowledgment

This work was conducted with the sponsorship of the U.S. Department of Energy under Contract No. DE-AC18-80FC10159. The authors wish to acknowledge the valuable assistance of S. Benson, D. Rindt, and A. Severson (UNDERC) in obtaining the analytical data, and of L. Delaney, T. Kuhlman, and C. Matone (BCRNL) in conducting the slag viscosity measurements. Thanks are also due to J. Waldron of Commonwealth Edison for providing the Black Butte coal sample, and to Dr. K. Mills of the British National Physical Laboratory for furnishing information on the IRSID and Urbain correlations.

Literature Cited

1. Watt, J.D.; Fereday, F. J. Inst. Fuel 1969, 42, 131-4.
2. Barrick, S.M.; Moore, G.F. ASME Paper No. 76-Wa/Fu-3, 1976.
3. Hoy, H.R.; Roberts, A.G.; Wilkins, D.M. Instn. Gas Engrs. Publication 672, 1964.
4. Watt, J.D.; Fereday, F. J. Inst. Fuel 1969, 42, 99-103.
5. "Investigation of the Distribution of Minerals in Coals by Normative Analysis," U.S. Department of Energy Report FE-2494-TR-2, 1980.
6. "Development, Testing, and Evaluation of MHD Materials," National Bureau of Standards Quarterly Reports Nos. 1-5, Contract No. E(49-18)-1230, 1974-5.
7. Riboud, P.V.; Roux, Y.; Lucas, L.D.; Gaye, H. Fachberichte Hüttenpraxis Metallweiterverarbeitung 1981, 19 (10), 859-66.
8. Urbain, G.; Cambier, F.; Deletter, M.; Anseau, M.R. Trans. J. Br. Ceram. Soc. 1981, 80, 139-41.

RECEIVED March 12, 1984

ASPECTS OF REACTIVITY

Role of Exchangeable Cations in the Rapid Pyrolysis of Lignites

MARK E. MORGAN[1] and ROBERT G. JENKINS

Fuel Science Program, Department of Materials Science and Engineering, Pennsylvania State University, University Park, PA 16802

The role of exchangeable cations on the rapid pyrolysis of lignites in nitrogen, carbon dioxide and wet nitrogen has been studied. In order to increase heating and cooling rates and to decrease secondary reactions of the pyrolysis products, an entrained-flow dilute-phase reactor was utilized. The effect that these alkali and alkaline-earth metals have on pyrolysis has been gauged by measuring the total weight loss, tar release, decarboxylation, and kinetics of total weight loss and decarboxylation. Additionally, a complete characterization of the discrete mineral phases and exchangeable cations is presented. It has been found that the presence of cations drastically alters pyrolysis behavior. For example, it was found that the presence of cations results in a decrease in total weight loss from 50% (no metal cations) to 30% (metal cations present). The presence of cations results in a three-fold decrease in tar release and a decrease in the extent of decarboxylation. Furthermore, the weight loss of the raw lignite increases from 30% in nitrogen to 40% in carbon dioxide in less than 0.5 s while the behavior of the acid-washed coal is relatively unaffected.

Recently, research concerned with utilization of the vast reserves of American lignites has expanded greatly. Studies have shown that lignites react quite differently than coals of higher rank when subjected to utilization and conversion schemes. The behavior of lignites is believed to be greatly influenced by the inorganic constituents present. The most distinctive feature of the inorganic constituents of lignites is the large concentration of exchangeable metal

[1]Current address: Atlantic Research Corporation, Alexandria, VA 22314

0097-6156/84/0264-0213$06.00/0
© 1984 American Chemical Society

cations. These cations are mainly alkali and alkaline-earth metals
associated with the carboxyl groups present in lignites.

This investigation is concerned with the role of exchangeable
metal cations in coal pyrolysis. There are three main areas to be
addressed: characterization of the inorganic constituents present,
the role of metal cations during pyrolysis in nitrogen, and the role
of metal cations during pyrolysis/gasification in reactive atmospheres of carbon dioxide and wet nitrogen. In order to study the
role of metal cations, a number of gauges of pyrolysis are considered, including total weight loss, decarboxylation, tar release,
and kinetics of total weight loss and decarboxylation. It should
be mentioned that the work presented here is intended to be a summary of a more extensive study which will be reported elsewhere (1).

Previous researchers have been concerned with some of these
topics. Tyler and Schafer (2), Franklin (3), and Otake (4) studied
the role of exchangeable cations in total weight loss and tar yield
under a variety of reaction conditions. In general, they found that
replacing the metal cations with hydrogen results in an increase
in weight loss. Murray (5), and Schafer (6,7) studied the decarboxylation of low-rank coals in a fixed-bed reactor with slow heating
rates. They found that removing the metal cations results in an
increase in the amount of decarboxylation at any given temperature.

In this investigation, all of the topics previously listed were
considered for the pyrolysis of a Montana lignite in a dilute-phase
reactor which employs high heating rates. Under these conditions,
the importance of secondary reactions occurring to the primary volatile products is minimized. Thus, the role of metal cations during
lignite pyrolysis has been studied by considering a large variety
of gauges of pyrolysis behavior in a reactor which allows inspection
of primary devolatilization process under rapid heating rates.

Experimental

The entrained flow reactor utilized in this study is virtually the
same unit described by Scaroni et al. (8). In order to improve knowledge about, and control of, the time-temperature history of the coal
particles, modifications have been made to the reactor and to the
predictions of the time-temperature history of particles in the reactor (9,10).

A schematic diagram of the entrained flow reactor is shown in
Figure 1. At the top of the reactor, a screw feeder and semi-venturi
system is used to entrain the ground coal particles in the cold primary gas stream. The coal is then injected into the reactor where
it is entrained in, and heated by, the preheated secondary gas. The
pyrolyzing coal particles fall in a thin stream through the reactor
and are collected by a movable water-cooled collector probe. The
time which the particles spend in the reactor is controlled by moving
the collector probe up and down the reactor axis. The pyrolysis
reactions are rapidly quenched in the collector probe, and the particles are separated from the gas stream by a cyclone in the collection system.

One of the most important features of this reaction system is
its high heating and cooling rates (10^4-10^5 K/s). This feature results in excellent control over the time-temperature history of the
pyrolyzing coal particles. Another important feature is the fact

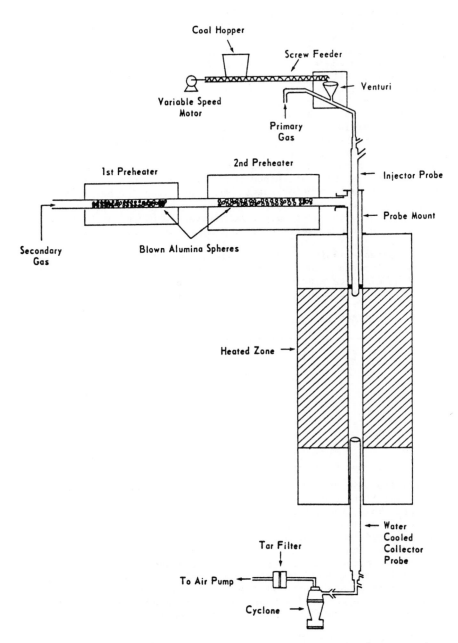

Figure 1. Schematic diagram of atmospheric pressure entrained
flow furnace.

that pyrolysis occurs in a very dilute phase. The dilution results
in a minimization of secondary reactions caused by interaction of
the pyrolysis products.

In addition to the equipment displayed, another probe was utilized for extended residence time studies. This probe consists of
a mullite tube with a ceramic crucible affixed at the top. This
probe was utilized to capture and hold particles in the reactor for
10 minutes so that pyrolysis could be completed.

In this paper, the use of three pyrolysis atmospheres will be
discussed: nitrogen, carbon dioxide and wet nitrogen. It should
be noted that the volumetric flow rates, and hence, gas velocities,
are the same for all four cases. Also, in each case, the primary
gas used to entrain the coal particles is helium. Helium was utilized because it was determined that the use of this gas results
in increased heat transfer rates (9). The amount of helium employed
accounts for about 3-4% (volume) of the total gases. The wet nitrogen referred to is water saturated nitrogen produced by bubbling
dry nitrogen through water at room temperature (i.e., 2.7% water
by volume).

A Montana lignite (Fort Union Seam) was utilized in this study.
The major inorganic constituents present in this lignite can be clas-
sified as discrete mineral phases and ion-exchangeable cations.
Table I shows the amount of the exchangeable cations present in the

TABLE I.   Cation Contents of Montana Lignite

Cation	10^{-4} g/g DICF Coal
Mg	61.6
Ca	218.5
Na	10.3
K	5.5
Ba	11.2
Sr	3.4
Total	310.5

raw lignite on a dry inorganic constituent free basis (DICF). This
basis takes into account the minerals plus the exchangeable cations
present. Thus, there are 0.082 g of minerals and 0.028 g of exchangeable cations per g of dry coal or 0.89 g DICF coal/g dry coal.
The concentration of cations was determined by extracting the coal
with 1N ammonium acetate, followed by analysis of the extract by
atomic absorption spectroscopy (11). The predominant cations present
are calcium and magnesium which is typical of an American lignite.
The concentration of carboxyl groups present is 3.1 mequiv/g DICF
(11). Of these carboxyl groups, it is estimated that 54% are exchanged with the metal cations present. The rest are in the acidic
form.

Characterization of discrete mineral phases present in the low
temperature ash (LTA) of the lignites was accomplished by a combination of x-ray diffraction and infrared spectroscopy. The procedure
for producing the LTA follows that described by Miller (12). The
percentage of calcite, quartz and pyrite are determined by x-ray

diffraction, while kaolinite and anhydrite are estimated by infrared spectroscopy (13). Table II shows the results of the analyses of the LTA from the ammonium acetate treated coal. This sample was utilized to prevent the formation of artifacts due to the presence

TABLE II. Mineral Matter Composition of Montana Lignite

Mineral	wt%/g Dry Mineral Matter
Kaolinite	41
Quartz	26
Anhydrite	4
Calcite	0
Other Clays	24
Others (by difference)	5

of metal cations during low-temperature ashing (11). As can be seen, the mineral matter is mainly comprised of clays and quartz. The mineral matter content of the raw coal is 8.2% on a dry coal basis.

In this study, three variations of the Montana lignite were utilized: raw lignite, acid-washed lignite and cation-loaded lignite. In the acid-washing procedure, the raw lignite is washed repeatedly with 0.1 N HCl solutions (11). After acid washing, the coal was stirred in boiling distilled water to remove excess HCl (14).

Cation loading was always performed on the acid-washed samples because determination of the extent of cation loading is facilitated by putting the carboxyl groups in the acid form. The conditions utilized for ion exchange can be seen in Table III; further details can be found elsewhere (9). All the samples were exchanged for 24 h,

TABLE III. Ion Exchange Conditions and Results

Sample	Exchange Media	Starting pH	g coal/l	m moles/g DICF Coal
Ca1	1 M Ca(OAc)$_2$	8.0	55	0.84
Ca2	.25 M Ca(OAc)$_2$ + .25 M CaCl$_2$	6.0	100	0.75
Ca3	.25 M Ca(OAc)$_2$	5.5	160	0.49
Ca4	1 M Ca(OAc)$_2$	8.0	55	1.11
Na	1 M Na(OAc)	8.0	55	0.94
Mg	1 M Mg(OAc)$_2$	8.0	55	0.71
Raw	--	---	---	0.94

with the exception of Ca3 which was exchanged for 96 h. Also shown in Table III are the results of the cation-loading preparations. As can be seen, there are four different loadings of calcium, ranging from 0.49 to 1.11 mequiv/g. The calcium loading of the raw lignite is closest approximated by the sample Ca1. Two samples were prepared by exchange with sodium and magnesium, respectively.

There are many ways in which the effects of pyrolysis may be gauged, one of the most informative of which is weight loss. Because

of the experimental conditions utilized, weight loss was determined
by using the ash-as-a-tracer method. This technique has been used
and evaluated previously by a number of investigators (8,15,16).

In addition, the decomposition of the carboxyl groups was
studied by measuring the carboxylate concentrations in the chars
after pyrolysis. The technique to determine the carboxyl group content is outlined in detail elsewhere (11), and is based on the work
of Schafer (17). There are three basic steps involved: acid washing, exchange with barium acetate and determination of the extent
of exchange.

The amount of tars released by the pyrolyzing coal was estimated
by collecting tars on the filter shown in Figure 1. It should be
mentioned that all the tars could not be collected in this manner
because some of them condense on the inside of the collector probe
and the associated tubing. However, this technique yields a qualitative gauge of the effect of ion-exchangeable cations on the amount
and chemical nature of the tars released.

Results and Discussion

Total Weight Loss. In Figure 2, the weight loss (at 1173 K, N_2)
versus time behavior for the raw, acid-washed and Cal-form samples
are displayed. As can be seen, the presence of metal cations dramatically affects the weight loss behavior in the entrained flow
reactor. All three samples undergo a period of rapid weight loss
followed by one of slow weight loss. In the case of the raw and
Cal-form samples, rapid weight loss lasts for about 0.15 s until
a value of about 30 wt% (DICF) is achieved. For the acid-washed
sample, the rapid weight loss is completed in 0.05 s and the final
weight loss value is 50 wt% (DICF). It is very interesting to note
the similarity between the behavior of the raw and Cal samples. This
similarity indicates the reversibility of the ion exchange treatments
in terms of coal pyrolysis behavior. This strongly suggests that
the difference in pyrolysis behavior between the metal cation and
acid-washed samples is mainly due to the presence of cations, and
not to the chemical treatments to which the coals were subjected.

While sample Cal contains about the same amount of metal cations
as the raw coal, it will be recalled that it was possible to vary
the calcium content from 0 to 1.11 mmoles/g DICF coal (Table III).
Pyrolysis of the four calcium-loaded and acid-washed samples showed
that there was a gradual and significant decrease in weight loss
as the calcium level was increased. The maximum weight loss at
1173 K within the reactor decreased from approximately 50% for the
acid-washed coal to 30% for the sample Ca4.

Other exchangeable cations are found in lignites; thus, a brief
study was made of the effects of sodium and magnesium on pyrolysis.
Results of this study are summarized in Table IV which lists weight
loss values at maximum residence time (~250 μs) for a number of different cation-loaded samples pyrolyzed at 1173 K. Firstly, it can
be seen that the presence of metal cations always results in a decrease in weight loss. This result is important in itself, in that
it reemphasizes the importance of cations in lignite pyrolysis. Secondly, information can be gained about the relative activity of the
various cations studied. If one compares the two divalent cations

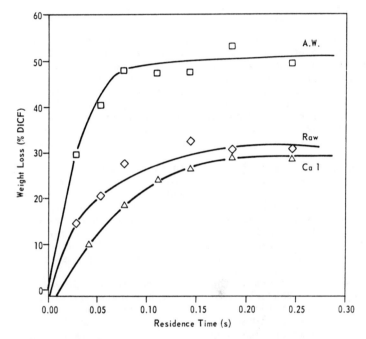

Figure 2.   Weight loss in entrained flow reactor in N_2 (1173 K) acid washed (A.W.) Montana lignite, raw and calcium exchanged (Cal) lignite, 270 x 400 mesh.

TABLE IV. Effect of Cations on Maximum Weight Loss in
Entrained Flow Reactor 1173 K, DICF Basis

Sample	Weight Loss
Raw	30.8
Ca1	28.7
Ca2	34.9
Mg	33.4
Na	38.7
Acid Washed	49.4

(calcium and magnesium), it can be seen that the elements have a
similar effect on a per mole basis. The samples Ca2 and Mg have
almost the same number of moles of cations, and both samples lose approximately the same weight. If one tries to assess the relative
activity of sodium, it can be seen that, in general, the sodium-
form coal undergoes a relatively large weight loss during pyrolysis.
The monovalent nature of sodium makes a comparison of the concentrations difficult. It contains the highest molar concentration of
cations, but they cover the least number of carboxyl groups.

Extended Residence Time Studies. In the entrained flow reactor,
coals undergo rapid weight loss followed by what appears to be a
region of little weight loss. However, although samples reach this
region of apparently constant weight loss value, no samples undergo
complete pyrolysis in the residence times available. Rather, this
"plateau" region of weight loss is a period of relatively slow decomposition. While it takes as little as 0.05 to 0.15 s to reach this
leveling off point, it can take on the order of minutes to complete
pyrolysis.

In order to determine the total pyrolysis yield, another technique was used. Basically, the approach involves capturing the pyrolyzing coal particles in a crucible at the maximum residence distance in the reactor. The captured samples are then held in the
reactor for 10 minutes to complete pyrolysis. It is thought that
this technique gives a reasonable estimation of the total pyrolysis
yield possible in an entrained flow reactor. The results of this
study can be seen in Table V in which weight loss values for extended

TABLE V. Extended Residence Time Runs in the Entrained
Flow Reactor 1173 K, DICF Basis

Sample	Weight Loss	A.S.T.M. Volatile Matter Content
Raw	54.0	43.7
Acid Washed	63.3	43.2
Ca1	46.1	43.0

residence time runs are listed. If one compares the weight loss of
these samples, it can be seen, once again, that the presence of metal

cations greatly reduces the amount of volatile material evolved.
It is interesting to compare these results with those gained by the
ASTM volatile matter test (18).   The ASTM volatile matter contents
display no effect of cation loading.   It will be recalled that this
test is performed in a fixed bed (crucible), which it is felt increases the amount of secondary char-forming reactions.   Thus, when
secondary char-forming reactions are increased, there is little difference in the weight loss values for pyrolysis.   However, decreasing the likelihood of these reactions (i.e., in the entrained flow
unit) leads to large differences in total weight loss.

Kinetics of Total Weight Loss.   The method chosen is a simple first
order Arrhenius treatment in which a single overall activation energy
is utilized.   The technique has been applied to the kinetics of the
initial weight loss, and therefore, the weight loss values were compared to the maximum weight loss in the entrained flow reactor.
Values of the apparent activation energy and preexponential factor
found in this study are listed in Table VI.   It is clear that there

TABLE VI.   Kinetic Parameters for Total Weight Loss

Sample	Activation Energy (kJ/mole)	Preexponential Factor (s^{-1})
Raw	58	8×10^3
Acid Washed	147	2×10^3
Cal	99	5×10^5

is a similarity in the behavior of the raw and calcium-form coals.
The acid-washed coal exhibits the largest activation energy, almost
three times larger than that found for the raw lignite and 50%
greater than that of the calcium form.   Again, it is obvious that
the absence of metal cations can have a profound effect on pyrolysis
kinetics and mechanisms.

Effect of Exchangeable Cations on Tar Release.   Table VII lists the
results of a study in which the quantity of tar, collected in a

TABLE VII.   Tars Released in Entrained Flow Reactor

Residence Time (s)	mg Tar/g Coal Fed	
	Raw Coal	Acid-Washed Coal
0.042	3	10
0.078	9	30
0.112	3	48

filter in the outlet gas stream, was measured for the raw and acid-
washed samples at 1173 K at three residence times.   At each residence
time the amount of tar released by the acid-washed coal was significantly greater than those released by the raw coals.   This trend
is in the same direction as the weight loss data.   That is, increases

in total weight loss when metal cations are removed are also accompanied by increases in tar yield. Tar samples (1173 K, 0.078 s) were analyzed by Fourier transform infrared spectroscopy. Spectra were determined using of KBr pellets in the manner described by Painter et al. (19). Spectra from the raw and the acid-washed coal tars were recorded as well as the "difference spectrum" from the two samples. Inspection of these spectra leads to the conclusion that the raw coal tars contain three times the quantity of aliphatic hydrogen as do the tars from the acid-washed coal.

Carboxyl Group Decomposition. Decomposition of the carboxyl groups was followed by measuring the quantity of carboxyl groups in the parent coal and in the resulting chars. It should be pointed out that data on the decomposition of carboxyl groups enables one to study the behavior of a single species during pyrolysis, thus yielding information that cannot be extracted from overall weight loss data. Also, it should be noted that since there are 3.1 meq/g DICF of carboxyl groups on this lignite, the decomposition of this species can account for a weight loss of up to 14% of the lignite.

Results from this study can be seen in Figure 3 in which the quantities of carboxyl groups remaining in the raw lignite as a function of residence time at 1173 K are shown. The loss of the carboxyl groups is very similar to the total weight-loss behavior presented in Figure 1. That is, there is a very rapid loss of carboxyl groups followed by a region of slow decomposition. It is very interesting to note that both the raw and Cal-form samples appear to complete decarboxylation at about 2.6 meq/g DICF while the acid-washed sample releases all of the 3.1 meq/g DICF present. This strongly suggests that the presence of metal cations retards carboxyl group decomposition. The fact that the raw and Cal samples decompose to the same extent is probably related to the fact that these samples contain essentially the same amount of cations.

Although previous researchers have studied the effects of cations on the pyrolysis of carboxyl groups, most of the work has been concerned with weight loss as a function of temperature and/or the evolution of carbon oxides. Little work has been concerned with direct determination of the kinetics of decarboxylation. Table VIII

TABLE VIII. Kinetic Parameters for Decarboxylation

Sample	Activation Energy (kJ/mole)	Preexponential Factor (s^{-1})
Raw	106	4×10^5
Acid Washed	80	8×10^4

lists results of a first-order kinetic analysis of decarboxylation of the raw and acid-washed forms of the lignite. To perform this analysis, decarboxylation information at 1173 K, 1073 K and 973 K was utilized. The activation energies for decarboxylation are somewhat different for the two samples studied; that found for the raw lignite is about 20% greater than that calculated for the acid-washed coal. However, the values are quite similar when compared to the

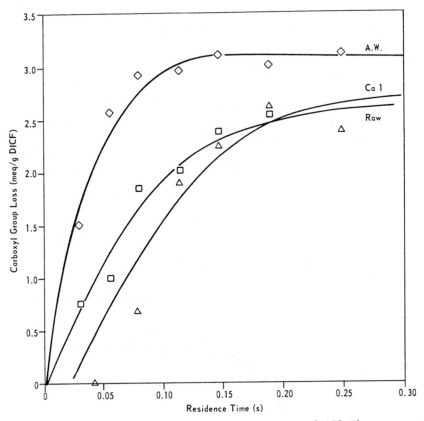

Figure 3. Decomposition of carboxyl groups (1173 K).

spread of activation energies found for the overall weight loss data
(Table VI).

<u>Pyrolysis in Wet Nitrogen and Carbon Dioxide</u>. The results of pyro-
lyzing the lignite in these atmospheres 'can be seen in Figures 4
and 5. In Figure 4, the results for pyrolysis of the raw coal in
nitrogen, carbon dioxide and wet nitrogen are presented. Here it
can be seen that the initial rates of weight loss in nitrogen and
carbon dioxide are similar and greater than in wet nitrogen. This
indicates a slower heating rate of the particles when wet nitrogen
is the secondary gas. Secondly, it is observed that a plateau region
is reached for all three gases and that the leveling off occurs at
about 31% in nitrogen, 34% in wet nitrogen and 39% in carbon dioxide
(DICF).

     Results for pyrolysis of the acid-washed samples in the three
atmospheres can be seen in Figure 5. In the case of this sample,
the same effect of atmosphere on initial weight loss can be observed.
That is, initial rates of weight loss in nitrogen and carbon dioxide
are similar and greater than the rates in wet nitrogen. However,
it can be seen that there is little, if any, effect of the atmospheres on final weight loss.

     If one compares Figures 4 and 5, an additional interesting re-
sult can be found. Although there is a significant increase in weight
loss when the raw coal is pyrolyzed in the reactive gases, the
largest weight loss for the raw coal is about 40% while all of the
acid-washed samples lose 50%.

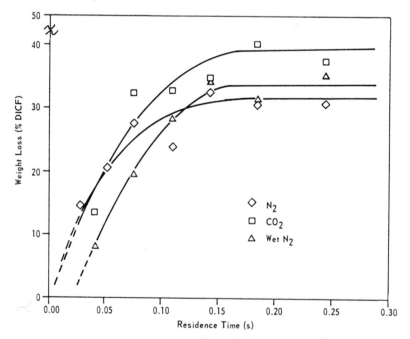

Figure 4. Weight loss in entrained flow reactor in N_2, CO_2 and
wet N_2 raw Montana lignite, 270 x 400 mesh.

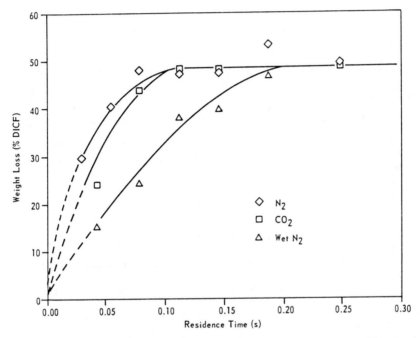

Figure 5. Weight loss in entrained flow reactor in N_2, CO_2, and wet N_2 acid washed Montana lignite, 270 x 400 mesh.

Conclusions

It must be concluded that the presence of metal cations in lignites dramatically affects pyrolysis behavior under rapid heating conditions. The presence of metal cations is important in evolution of volatile matter in nitrogen and in potentially reactive gases. These cations influence the amount, nature and rate of volatile release. Furthermore, it is suggested that these data have significant importance in gasification and combustion of lignites.

Acknowledgments

This study was supported by USDOE under Contract No. DE-AC01-79ET14882 and, in part, by The Pennsylvania State University Cooperative Program in Coal Research. We wish to thank Dr. Spackman, Director of the Coal Research Section (PSU) for supplying the lignite.

Literature Cited

1. Morgan, M. E. and Jenkins, R. G., submitted to Fuel for publication.
2. Tyler, R. J. and Schafer, H. N. S., Fuel, 1980, (59), 487.
3. Franklin, H. D., Cosway, R. G., Peters, W. A., and Howard, J. B., Ind. Eng. Chem. Process Des. Dev., 1983, (22), 39.
4. Otake, Y., M.S. Thesis, The Pennsylvania State University, 1982.
5. Murray, J. B., Fuel, 1978, (57), 605.
6. Schafer, H. N. S., Fuel, 1979, (58), 667.
7. Schafer, H. N. S., Fuel, 1979, (58), 673.
8. Scaroni, A. W., Walker, P. L., Jr., and Essenhigh, R. H., Fuel 1981, (60), 71.
9. Morgan, M. E., Ph.D. Thesis, The Pennsylvania State University, 1983.
10. Maloney, D. J., Ph.D. Thesis, The Pennsylvania State University, 1983.
11. Morgan, M. E., Jenkins, R. G., and Walker, P. L., Jr., Fuel, 1981, (6), 189.
12. Miller, R., Ph.D. Thesis, The Pennsylvania State University, 1978.
13. Jenkins, R. G. and Walker, P. L., Jr., in Analytical Methods for Coal and Coal Products, Vol. II (Clarence Karr, Jr., Editor), Academic Press, New York, 1964.
14. Dee, G. S., Tumuluri, S., and Lahari, K. C., Fuel, 1971, (50), 272.
15. Anthony, D. B. and Howard, J. B., AIChE Journal, 1976, (22), 625.
16. Badzioch, S. and Hawksley, P. G. W., Ind. Eng. Chem. Process Des. Dev., 1970, (9), 521.
17. Schafer, H. N. S., Fuel, 1970, (49), 197.
18. ASTM, Annual Book of ASTM Standards, Part 26, American Society for Testing and Materials, Philadelphia, PA, 1978.
19. Painter, P. C., Kuehn, D. W., Snyder, R. W., and Davis, A., Fuel, 1982, (61), 691.

RECEIVED March 12, 1984

Low-Rank Coal Hydropyrolysis

C. S. WEN and T. P. KOBYLINSKI

Gulf Research and Development Company, Pittsburgh, PA 15230

Recently, increased attention is being directed toward the pyrolysis route of processing coal to produce liquid and gaseous fuels, particularly when coupled to the use of char by-product for power generation.(1) In view of the increased interest in coal pyrolysis, a better understanding of the thermal response of coals as they are heated under various conditions is needed.

Conventional liquefaction processes developed for Eastern bituminous coals might not be the best choice for Western low-rank coals because of the substantial property and structural differences between them. In general, low-rank coals are more susceptible to reaction with H_2, CO, or H_2S.

Coal hydropyrolysis is defined as pyrolysis under hydrogen pressure and involves the thermal decomposition of coal macerals followed by evolution and cracking of volatiles in the hydrogen. It is generally agreed that the presence of hydrogen during the pyrolysis increases overall coal conversion.(2,3)

In the present study we investigated pyrolysis of various ranks of coals under different gaseous environments. Low-rank coals such as Wyoming subbituminous coal and North Dakota lignite were pyrolyzed and their results were compared with Kentucky and Illinois bituminous coals.

Experimental

The coals used in this study included Wyoming subbituminous coal, North Dakota lignite, Kentucky bituminous coal, and Illinois No. 6 bituminous coal. Proximate and ultimate analyses of the coals studied are given in Table I.

All pyrolysis experiments were carried out in the thermo-gravimetric apparatus (TGA) having a pressure capacity of up to 1000 psi. A schematic of the experimental unit is shown in Figure 1. It consists of the DuPont 1090 Thermal Analyzer and the micro-balance reactor. The latter was enclosed inside a pressure vessel with a controlled temperature programmer and a computer data storage system. The pressure vessel was custom manufactured by Autoclave Engineers. A similar set-up was used previously by others.(4)

0097–6156/84/0264–0227$06.00/0
© 1984 American Chemical Society

Table I. Coal Analysis

	Wyoming Subbituminous Coal		North Dakota Lignite		Kentucky Bituminous Coal		Illinois Bituminous Coal	
	As Received	Dry Basis	As Received	Dry Basis	As Received	Dry Basis	As Received	Dry Basis
Proximate Analysis, wt%								
Moisture	6.1	--	21.2	--	3.0	--	3.9	--
Volatile	39.9	42.5	35.3	44.7	39.8	41.0	38.1	39.6
Fixed Carbon	45.6	48.6	35.9	45.6	48.6	50.1	46.8	48.7
Ash	8.4	8.9	7.6	9.7	8.6	8.9	11.2	11.7
Ultimate Analysis, wt%								
Moisture	6.1	--	21.2	--	3.0	--	3.9	--
Carbon	64.6	68.8	50.5	64.1	70.3	72.5	67.5	70.3
Hydrogen	4.3	4.6	3.5	4.4	4.8	5.0	4.5	4.7
Nitrogen	0.9	1.0	1.1	1.4	1.2	1.2	1.3	1.4
Sulfur	0.5	0.5	0.5	0.6	3.0	3.1	3.4	3.5
Ash	8.4	8.9	7.6	9.7	8.6	8.9	11.2	11.7
Oxygen (diff.)	15.1	16.1	15.6	19.7	9.1	9.4	8.2	8.4

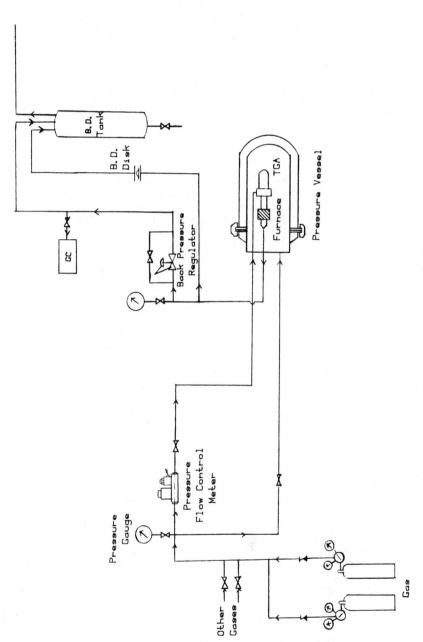

Figure 1.   Schematic setup of the pressure pyrolysis apparatus.

A chromel-alumel thermocouple was set in close proximity to the
sample inside a reactor. The reactor was made of a quartz tube
which was surrounded by a tubular furnace. In a typical coal
pyrolysis run, the coal sample (20-30 mg) was placed in a platinum
boat which was suspended from the quartz beam of the TGA balance.
The coal particle size used was 100-200 mesh. Samples were heated
to desired temperatures at linear heating rates or heated iso-
thermally under various gaseous environments.
    Fourier transform infrared spectroscopy (FTIR) was also used to
monitor the degree of pyrolysis for various samples at different
temperatures. The KBr (potassium bromide) pellet of sample was
prepared for FTIR analysis.

Results and Discussion

Typical thermograms of Wyoming coal under hydrogen pressure are
given in Figure 2. The TGA and the weight-loss rate thermograms
show a major weight loss at temperatures ranging from 350-600°C. A
secondary hydropyrolysis peak occurs below 600°C most likely due to
the gas releases from the further decomposition of coal. Figure 3
shows a comparison of derivative thermograms for four different rank
coals. The differences of devolatilization rate are not large at
temperatures up to 500°C; however, above 500°C, the secondary
hydropyrolysis peak of low-rank coal becomes dominant.
    The effect of hydrogen on coal pyrolysis can further be
illustrated by Figure 4, where we compared derivative thermograms of
Wyoming coal pyrolyzed at 200 psig of N_2 and 200 psig of H_2 at the
same heating rates (20°C/min). The secondary hydropyrolysis peak
observed at 580°C in the H_2 run was absent in the N_2 atmosphere.
    The influence of heating rate on coal hydropyrolysis was
studied over a range of 5-100°C/min. As shown in Figure 5, two
peaks were observed, the first of which we call the primary
volatilization, and the second, characteristic of local hydropyrolysis. The first peak increased rapidly with the increase of the
heating rate. The second characteristic peak becomes relatively
dominant at lower heating rates. Although the particle size range
(100-200 mesh) used was not small enough to eliminate particle size
as a factor in the pyrolysis, it seems that the hydropyrolysis peak
is favored by slow heating rates, indicating a heat transfer and/or
mass transfer limitations within the secondary hydropyrolysis
region. In contrast, coal pyrolyzed under an inert nitrogen
atmosphere results in an increase of weight-loss rate with increasing heating rates, but the shape of the curves remains the same
(Figure 6).
    FTIR spectra of the original coal and char from pyrolyzing
Wyoming coal under H_2 pressure at various temperatures are shown in
Figure 7 (only wave numbers between 1700 and 400 cm^{-1} were shown
here for comparison). The strongest absorption band located at
1600 cm^{-1} begins to decrease in intensity at 380°C and shows a
marked decrease at 470°C. This band has been assigned to aromatic
ring C-C vibration associated with phenolic/phenoxy groups.(5)
Similarly, the aromatic-oxygen vibration band near 1260 cm^{-1} shows
equivalent decreases. These observations are in good agreement with
the TGA weight-loss curve (as shown in Figure 2) where the onset

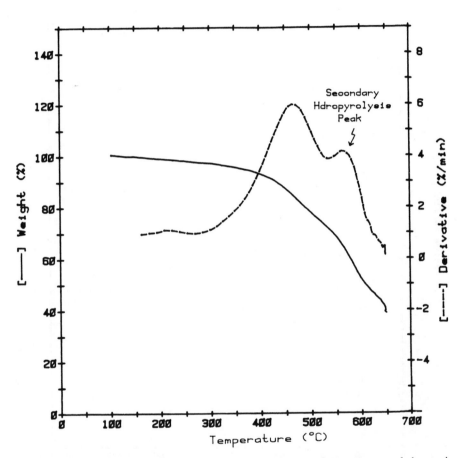

Figure 2. Typical pressure TGA thermograms of Wyoming coal heated at 50 °C/min and 500 psig H_2.

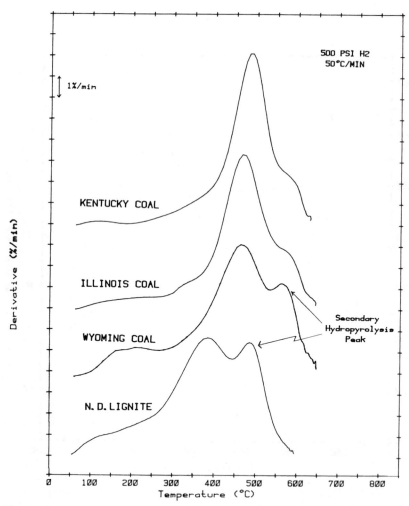

Figure 3. A comparison of hydropyrolysis derivative thermograms for various coals at 50° C/min and 500 psig H_2.

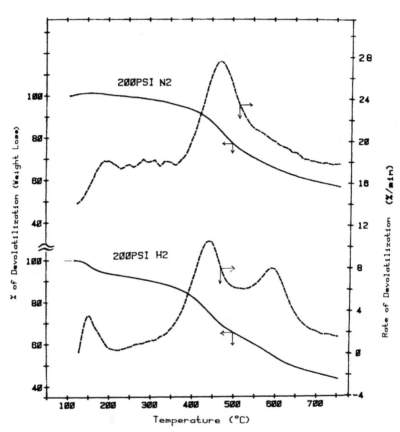

Figure 4. A comparison of thermograms between Wyoming coal pyrolyzed in 200 psig N_2 and 200 psig H_2 at 20 °C/min.

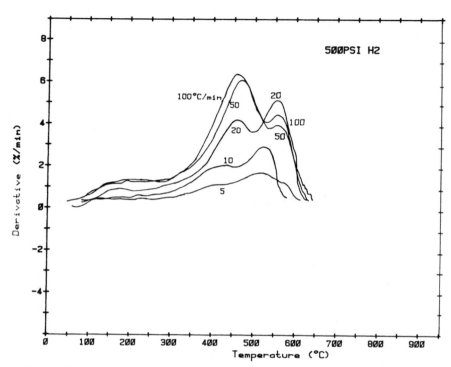

Figure 5. The effect of heating rate on devolatilization rate
of Wyoming coal hydropyrolysis.

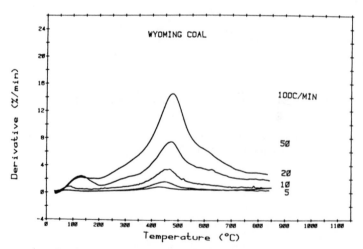

Figure 6. Derivative thermograms of coal pyrolysis at different
heating rates under 1 atm N_2.

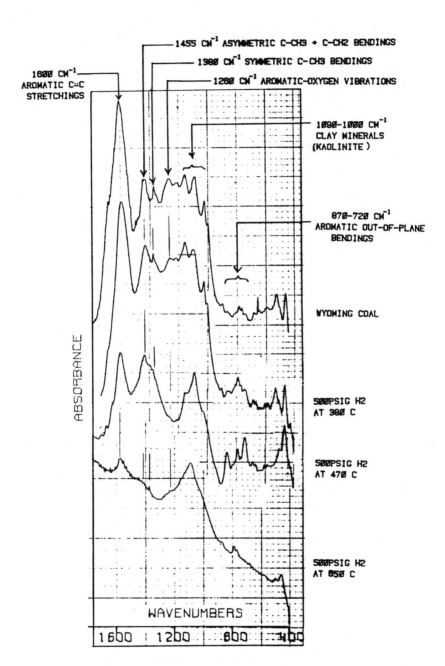

Figure 7.   FT–IR monitor of Wyoming coal pyrolysis under 500 psig
H_2 at various temperatures.

temperature of devolatilization occurs at around 350°C and the peak temperature of volatilization located at 467°C, suggesting the initial coal devolatilization involving the decomposition of oxygen-containing groups.

The absorption bands between 720 to 870 cm^{-1} (which arise from the out-of-plane aromatic CH vibrations) increase markedly at 470°C, indicating a growth in size of aromatic clusters in the reacted coal. These aromatic out-of-plane peaks disappear as the secondary hydropyrolysis peak (shown in the TGA weight-loss curve) takes place. Therefore, at 650°C, most organic absorption bands were diminished except for a broad band ranging from 1000-1090 cm^{-1} due to clay mineral absorption.(7) An increase in the structureless background absorption is observed, suggesting a growth of graphitization in the residual coal.

The kinetics of coal pyrolysis are complicated because of the numerous components or species which are simultaneously pyrolyzed and decomposed. The chemical expression in coal pyrolysis is represented briefly by an overall reaction:

$$W \xrightarrow{k} x V + (1-x) S \qquad (1)$$

where W, coal; V, total volatile yield; S, char; and rate constant $k = A \exp(-E/RT)$; A, E, and x, frequency factor, activation energy, and stoichiometric coefficient, respectively; R, gas constant; and T, temperature of coal particle.

At any temperature, T, the rate of devolatilization of coal may be written as:

$$\frac{-dw}{dt} = k\, w^{n} \qquad (2)$$

or

$$\frac{-dw}{dt} = A \exp(-E/RT)\, w^{n} \qquad (3)$$

For measuring kinetic parameters, we treated data following the procedure of Coats and Redfern (8) and Mickelson and Einhorn (9) by taking logarithms from Equation (3) at temperatures, T_i and T_j:

$$\ln\left(-dw/dt_j / -dw/dt_i\right) = \frac{E}{R}\left(T_j - T_i\right)/T_i T_j - n \ln(w_i/w_j) \qquad (4)$$

At a constant ratio of w_i/w_j, a plot of

$$\ln\left(-dw/dt_j / -dw/dt_i\right) \text{ vs } \left(T_j - T_i\right)/T_i T_j \qquad (5)$$

should show the straight line plot with the data points where E and n can be determined. The frequency factor can be further computed, based on the known values of n and E and on an approximate integral solution to Equation (3).

The kinetic parameters for four different coals heated under H_2 pressure are presented in Table II. A reaction order equal 2.3 to 2.9 was observed for the primary hydropyrolysis peaks. A high reaction order was obtained for the secondary reaction peak under hydrogen pressure. The kinetic parameters for four different coals heated under N_2 atmosphere compared with data presented in the literature (6,10-12) are listed in Table III.

Kinetic parameters for n (reaction order) and E (activation energy) in the literature show significant variation for different techniques and coal (Table III). By considering the complexity of coal thermal degradation, many authors have contended that a simple, first-order reaction is inadequate. Wiser et al. (10) found that n=2 gave the best fit to their data, while Skylar et al. (11) observed that values of n above 2 were required to fit nonisothermal devolatilization data for different coals. The kinetic parameters obtained in this study also show a non-integral reaction order. The thermal decomposition of coals are complex because of the numerous components or species which are simultaneously decomposed and recondensed.

Figure 8 demonstrates the influence of CO on devolatilization for different ranks of coal. For the higher rank coals (i.e., Kentucky and Illinois bituminous coals), pyrolysis in the presence of CO plus H_2 or H_2 alone follows the same path. However, lignite showed a marked increase of pyrolysis rate in the run where CO was added. A higher content of reactive oxygenated bonds (i.e., carboxyl or ether linkages) in low-rank coals could be the reason for the high reactivity in the presence of CO. The kinetic parameters determined for coal pyrolyzed in syngas (CO/H_2 mixture) are listed in Table IV. As shown in the table, a high reaction order was obtained for the H_2 and CO/H_2 runs, particularly for the low-rank coals which showed a secondary reaction occurring at temperatures above 500°C.

Conclusions

Laboratory microscale studies have demonstrated that the coal pyrolysis in a hydrogen atmosphere gave higher degree of devolatilization in low-rank coals than pyrolysis in an inert atmosphere. In hydrogen atmosphere two distinct steps in coal devolatilization were observed as shown by the double peak of the devolatilization rate. Only one step was observed under nitrogen atmosphere. In a comparison of kinetic parameters, a high reaction order and a low activation energy were also obtained in the coal hydropyrolysis. Application of data and observations from this study could lead to a better understanding of chemical and physical changes during the coal hydropyrolysis and seek alternative coal conversion routes.

Table II. Kinetic Parameters of Coal Hydropyrolysis[a]

Coal	Reaction Order		Activation Energy (kcal/mole)		Frequency Factor (min^{-1})	
	1st Peak	2nd Peak[b]	1st Peak	2nd Peak[b]	1st Peak	2nd Peak[b]
Wyoming	2.5	3.4	18.9	64.5	1.6×10^5	1.3×10^{15}
North Dakota Lignite	2.9	4.7	23.9	70.4	4.2×10^7	1.2×10^{18}
Kentucky	2.3	--	23.4	--	3.1×10^6	--
Illinois	2.4	--	29.2	--	1.9×10^8	--

a Samples were heated at 50°C/min under 500 psig H_2.
b Hydropyrolysis characteristic peak occurred in low-rank coals.

Table III. A Comparison of Kinetic Parameters in Coal Pyrolysis

Investigators	Coal	Reaction Order	Activation Energy (kcal/mole)	Frequency Factor (min^{-1})	Reference
Wiser et al.	Utah Bituminous	2	15.0	2.9×10^3	(10)
Stone et al.	Pittsburgh Seam Bituminous	1^a	27.3	3.2×10^8	(12)
Ciuryla et al.	Pittsburgh Seam Bituminous	1^a	39.5	1.7×10^{11}	(6)
	North Dakota Lignite	1^a	53.6	1.7×10^{15}	(6)
	Illinois Bituminous	1^a	52.3	1.7×10^{15}	(6)
Skylar et al.	Soviet Coal	2.3	10.0	1.3×10^5	(11)
	Soviet Gas Coal	2.1	14.6	5.1×10^6	(11)
This Work[b]	Wyoming Subbituminous	2.1	25.1	6.1×10^5	
	North Dakota Lignite	2.2	23.2	1.2×10^6	
	Kentucky Bituminous	1.9	26.9	4.2×10^8	
	Illinois Bituminous	1.8	30.9	4.8×10^8	

[a] Based on a series of first-order reactions.
[b] Samples were heated at 50°C/min under 50 cc/min ambient N_2 flow.

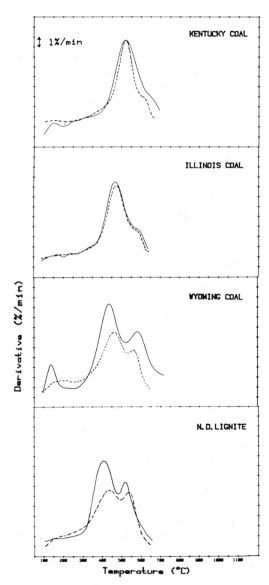

Figure 8. The influence of CO on coal devolatilization, where
—— is 500 psig CO/H_2, and --- is 500 psig H_2.

Table IV. Kinetic Parameters of Coal Pyrolysis Under Syngas[a]

Coal	Reaction Order		Activation Energy (kcal/mole)		Frequency Factor (min^{-1})	
	1st Peak	2nd Peak	1st Peak	2nd Peak	1st Peak	2nd Peak
Wyoming Coal	2.8	3.6	23.1	57.5	1.1×10^{7}	7.1×10^{14}
North Dakota Lignite	2.6	2.8	20.8	32.6	2.6×10^{6}	1.8×10^{12}
Kentucky Coal	2.3	--	22.9	--	2.3×10^{6}	--
Illinois Coal	2.3	--	25.4	--	3.5×10^{7}	--

[a] Samples were heated at 50°C/min under 500 psig H_2O/CO (3/1 mole ratio).

Literature Cited

1. Parker, H. W. "Liquid Synfuels Via Pyrolysis of Coal in Association with Electric Power Generation," Energy Progress 1982, 2(1), 4.
2. Belt, R. J.; Bissett, L. A. "Assessment of Flash Pyrolysis and Hydropyrolysis," DOE Report METC/RI-79/2, 1978.
3. Arendt, P.; Van Heek, K. H. Fuel 1981, 60(9), 779.
4. Dobner, S.; Kan, G.; Graff, R. A.; Squires, A. M. "Modification of a Thermobalance for High Pressure Process Studies," AIChE Meeting, Los Angeles, November 16-20, 1975.
5. Speight, J. G. "Assessment of Structures in Coal by Spectroscopic Techniques"; "In Analytical Methods for Coal and Coal Products"; Karr, C., Jr., Ed.; Academic Press: New York, 1978; Vol II, Chapter 22, pp. 75-101.
6. Ciuryla, V. T.; Weimer, R. F.; Bivans, D. A.; Motika, S. A. Fuel 1979, 58(10), 748.
7. Painter, P. C.; Coleman, M. M. "The Application of Fourier Transform Infrared Spectroscopy to the Characterization of Coal and Its Derived Products," DIGILAB FTS/IR Notes No. 31, January 1980.
8. Coats, A. W.; Redfern, J. P. "Kinetic Parameters from Thermogravimetric Data," Nature 1964, 201, 68.
9. Mickelson, R. W.; Einhorn, I. N. Thermochim. Acta 1970, 1, 147.
10. Wiser, W. H. Fuel 1968, 47, 475, and Wiser, W. H.; Hill, G. R.; Kertamus, N. J. I&EC Process Design and Develop. 1967, 6(1), 133.
11. Skylar, M. G.; Shustikov, V. I.; Virozub, I. V. Intern. Chem. Eng. 1969, 9, 595.
12. Stone, H. N.; Batchelor, J. D.; Johnstone, H. F. I&EC 1954, 46, 274.

RECEIVED March 5, 1984

Combustion Reactivity of Chars from Australian Subbituminous Coals

B. C. YOUNG

Commonwealth Scientific and Industrial Organisation, Division of Fossil Fuels, PO Box 136, North Ryde, N.S.W. 2113, Australia

The reactivities to oxygen of sized fractions of chars produced by flash pyrolysis from Millmerran subbituminous coal at 540, ~600 and ~790°C and from Wandoan subbituminous coal at 550°C have been determined. The reactivity of the Wandoan char is about twice the reactivities of the Millmerran chars. The burning rates of the four chars show a half-order dependence on oxygen partial pressure, and activation energies in the region of 20 kcal/mol. This latter fact, together with observed changes in particle size and density with burn-off, shows that reaction rates are controlled by the combined effects of pore diffusion and chemical reaction. When the reactivities of the subbituminous coal chars are compared with those for other carbons, it can be seen that they are similar to those for brown coal char, bituminous coal char, anthracite and semi-anthracite but much higher than the reactivity of petroleum coke.

In order to advance the scientific understanding and technological application of pulverized-coal combustion, staff of the CSIRO Division of Fossil Fuels (and its antecedents) have been actively engaged in studies of the combustion kinetics of coal chars and related materials for at least 15 years. Early studies included an investigation of a char produced from a Victorian brown coal (1) which was found to have a half-order dependence on oxygen concentration. This finding is in contrast to the combustion behavior reported for other coal chars and petroleum coke, where a first order dependence was assumed or found to prevail (2-7). A subsequent investigation of the combustion rate of petroleum coke over an extended range of oxygen concentration (8) also revealed a half-order dependence.

More recently, the CSIRO work has included studies of chars produced from the flash pyrolysis of subbituminous coals. This work has formed part of a major project to develop the flash pyrolysis process of converting coal into oil (9) in which pulverized coal

0097-6156/84/0264-0243$06.00/0
© 1984 American Chemical Society

particles are rapidly heated in a fluidized sand bed for short ($\sim$ 1 s) contact times to give tar, gas and char. The char, which accounts for about 50% of the product yield is used in part to supply process heat, the remainder being available as a fuel for a power plant.

In this paper the combustion reactivities of four flash pyrolysis chars are compared with the results for chars produced from low and high-rank coals under conditions simulating pulverized-coal combustion, for anthracite and semianthracite, and petroleum coke. Reactivity is expressed as the rate of combustion of carbon per unit external surface area of the particle, with due correction being made for the effect of mass transfer of oxygen to the particle.

Theory

The rate, ρ, at which a carbon particle burns can be expressed by the following relation (10):

$$\rho = R_c \, [p_g \, (1-\chi)]^n \qquad g/cm^2s \qquad (1)$$

where p_g is the oxygen partial pressure in the bulk gas, and n the apparent or observed order of reaction with respect to p_g. R_c is a rate coefficient which incorporates the effects of oxygen diffusion into the pore structure of the particle and the rate of reaction of oxygen on the pore wall. χ is the ratio ρ/ρ_m, where χ is related to n and p_g, as follows (1, 11)

$$\chi/(1-\chi)^n = R_c p_g{}^n/\rho_m \qquad (2)$$

and for low values of χ ($\ll$1)

$$\chi_1/\chi_2 \; \simeq \; (p_{g,1}/p_{g,2})^{n-1} \qquad (3)$$

where the subscripts 1 and 2 denote different values of p_g.

ρ_m is the burning rate under mass transfer control, exhibited at very high values of R_c. It is the maximum possible burning rate, and is given by (5,12):

$$\rho_m = 5.61 \times 10^{-6} p_g \, [(T_P + T_G)/2]^{0.75}/d \qquad g/cm^2s \qquad (4)$$

where the numerical coefficient for the stoichiometry, $C + 1/2 \, O_2 \rightarrow CO$, incorporates the diffusion coefficient for oxygen in nitrogen. T_P and T_G are the particle and gas temperatures respectively, and d is the particle size.

It is convenient to express R_c in Arrhenius form, namely

$$R_c = A \exp \, [-E/RT_p] \qquad g/ \{ \, cm^2s \, (atm \, O_2)^n \, \} \qquad (5)$$

T_P was calculated from the steady-state heat balance (confirmed by earlier measurements (13)) as shown below:

$$\rho H - 2k \, (T_P-T_G)/d - \varepsilon\sigma_B(T_P{}^4-T_W{}^4) = 0 \qquad (6)$$

The heat of reaction, H, for carbon oxidation to carbon monoxide is 2340 cal/g. k is the gas thermal conductivity evaluated at the arithmetic mean of particle and gas temperatures; ϵ is the particle emissivity and assumed here to be unity; and σ_B is the Stefan-Boltzmann constant. T_W is the temperature of the reactor wall.

Changes in the particle structure have a strong effect on combustion behavior, influencing the particle temperature, mass transfer and pore diffusion rates, and consequently the rate-control regime of the process (5). The changes in size and density of particles that have a homogeneous pore structure and small pore sizes (relative to particle size) are related to fractional burn-off, u, by

$$d^3\sigma = d_o{}^3\sigma_o \ (1-u) \tag{7}$$

where σ is the particle density. The subscript o refers to initial conditions.

Two limiting cases can be evaluated (5):

$$d = d_o \ (1-u)^{1/3} \ \text{(particles burning at constant density)} \tag{8}$$
$$\sigma = \sigma_o \ (1-u) \ \text{(particles burning at constant size)} \tag{9}$$

In the case of particles burning at constant density, the reaction is confined to the exterior surface as burn-off progresses. For particles burning at constant size, it is assumed that complete penetration of oxygen into the pores occurs and the reaction proceeds uniformly through the entire particle during burn-off. However, it is also possible that the burning particles show behavior intermediate between the above cases, i.e. the pore structure may not be completely penetrated by oxygen. The adequacy or otherwise of these descriptions to the chars will be seen later.

Experimental Procedure

The chars were produced in a process development unit scale fluidized-bed flash pyrolyser (9) from pulverized Millmerran and Wandoan subbituminous coals from Queensland, Australia. Millmerran coal was pyrolyzed at various temperatures between 500 and 800°C and Wandoan coal at 550°C. The resulting chars were screened to yield a size fraction with a median mass size around 80 μm. The properties of the chars are presented in Table I.

Rate measurements were carried out in a tubular entrainment flow reactor illustrated schematically in Figure 1. The reactor was fed with a metered supply of size-graded char particles in a metered supply of hot gas which was generated by injecting the combustion products of an acetylene-air or acetylene-oxygen-nitrogen flame into a mixture of nitrogen and oxygen. Gas temperatures employed in the char studies were in the range 830 to 1350 K, with initial oxygen partial pressures ranging from 0.04 to 0.25 atm and the total pressure being close to atmospheric. The burn-out of the particles was followed by measuring the progressive decrease in oxygen and increase in carbon dioxide concentrations at the five ports along the reactor. The reaction rate, ρ (grams of carbon burned per square

Table I.   Properties of the Chars

	Millmerran					Wandoan
Pyrolysis temperature (°C)	540	585	610	780	800	550
Median mass size (μm)	85	76	90	70	88	76
Size below which 90% of material lies (μm)	100	95	118	92	98	96
Size below which 10% of material lies (μm)	66	52	61	47	74	49
Particle density (g/cm³)	0.88	0.80	0.78	0.99	0.78	1.05
Chemical analysis (% w/w, as received)						
Moisture	2.7	2.2	3.3	2.5	4.7	2.9
Ash	28.0	30.4	35.6	60.1	38.2	22.1
C	56.6	55.3	50.7	31.3	50.3	60.3
H	3.8	2.8	2.7	1.1	1.3	3.3
N + S + 0	8.9	9.3	7.7	5.0	5.5	11.4

centimetre of external surface of particle per second), was determined from the known inputs of solid and gaseous reactants, the distribution of gas and wall temperatures and the fractional burn-off.

When the rate is strongly influenced by mass transfer, i.e. when $\chi > 0.5$, values of R_c cannot be accurately calculated. Hence, the data presented here are for χ values less than 0.5.

Particle size and particle density measurements were made on partially burnt char particles which were collected in a cyclone separator at the exit of the reactor. Their burn-off was evaluated from a knowledge of char feedrate, gas flowrates and gas composition. The particle size of the collected material was determined by sieving; particle density was derived from measuring the bulk density of a bed of particles in a manner described by Field (2).

Results and Discussion

Figure 2 relates χ to temperature in Arrhenius form at various levels of p_g, for the Millmerran char produced at 585 and 610°C. Several important features are indicated by this plot:

(1) χ is strongly affected by temperature.
(2) At the higher temperatures a number of the values of χ exceed 0.5 (due to experimental uncertainty one value of χ exceeds the maximum theoretical value of unity).
(3) The general trend of the values suggests an inverse relationship between χ and the partial pressure of oxygen, consistent with a fractional reaction order (see Equation 3). The data lack sufficient precision to determine n from χ quantitatively. However, n can be estimated indirectly as shall be seen below.

Determination of the rate coefficient R_c as a function of particle temperature requires a knowledge of n. However, the

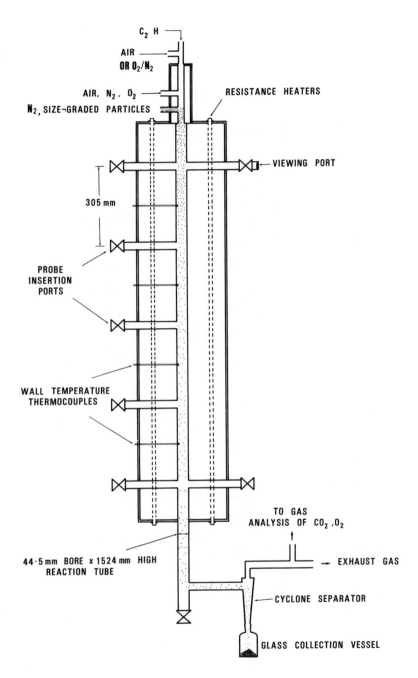

Figure 1.   Apparatus for pulverized-fuel combustion kinetics.

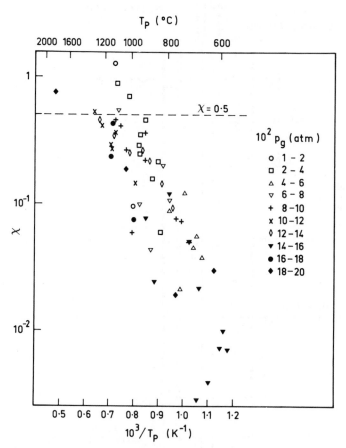

Figure 2.   Effect of particle temperature and oxygen partial pressure on χ for Millmerran char (585 and 610 °C).

experimental data do not allow an independent determination of n. Therefore R_c values were calculated at a series of assumed values of n in the range 0.1 to 1.0. The dependence of R_c on T_p was then determined in Arrhenius form. The optimum n value was ascertained applying a linear regression analysis of the data, to determine at which value of n the relationship between R_c and T_p displayed least scatter. As noted elsewhere (8), the best fit of the data is given by a minimum in the sums of the squares of the residuals.

For the present chars the least scatter in the rate data occurred at n = 0.5. This value of n is the same as that determined previously for Millmerran char (14), Yallourn brown coal char (1) and petroleum coke (8).

Rate data are shown in Arrhenius form in Figure 3 for Millmerran chars produced at 585 and 610°C. The data for these chars are combined into a single set (as are the data for the chars produced at 780 and 800°C presented in Table II).

Table II. Kinetic Data for the Combustion of Chars and Coke

Material	A g/ $\{cm^2s \, (atm \, O_2)^n\}$	E kcal/mol	n	Refer-ence
Millmerran char (540°C)	15.6	17.5	0.5	14
Millmerran char (585 and 610°C)	22.3	18.8	0.5	15
Millmerran char (780 and 800°C)	73.3	21.7	0.5	15
Wandoan char (550°C)	39.1	18.3	0.5	15
Yallourn brown coal char	9.3	16.2	0.5	1
New Zealand bituminous coal char	8	16.0	1.0	5
Anthracite and semianthracite	20.4	19.0	1.0	6
Petroleum coke	7.0	19.7	0.5	8

Values of A, E ánd n for the chars referred to in this study, together with those for other low-rank and some high-rank materials investigated at CSIRO, are given in Table II. The data have been used to compare the reactivity of the various materials as illustrated in Figure 4. Because the different materials have different values of n, the data are compared using a rate, ρ_c, calculated from the relation:

$$\rho_c = R_c \, p_g^{\ n} \qquad g/cm^2s \qquad (10)$$

where p_g is 1 atm.

Comparison of these data reveals the following notable features:

(1) The combustion rates of all the chars show a strong dependence on temperature, the values of the activation energies ranging from 16 to 22 kcal/mole.
(2) The activation energy increases and the reactivity of Millmerran chars decreases with increasing pyrolysis temperature.

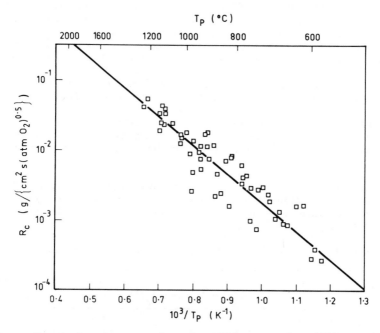

Figure 3.   Combustion rate data for Millmerran char (585 and 610 °C).

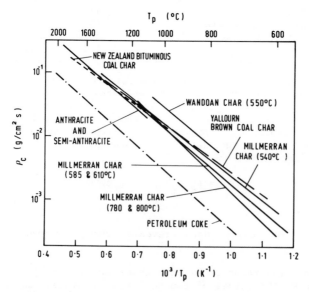

Figure 4.   Combustion rate data for chars and coke.

(3) The most reactive char is that produced from Wandoan coal at 550°C.

(4) At p_g = 1 atm the low-rank coal chars have reactivities which are three to ten times higher than the reactivity of petroleum coke. However, as some of the materials show apparent reaction orders of unity, and others of 0.5, relative reactivities at other values of p_g are different. For example, when p_g = 0.1 atm (a level found in practical flames) the reactivities of chars from brown and subbituminous coals are similar but are appreciably higher than the reactivities of the high-rank materials.

(5) The low-rank coal chars and petroleum coke have an apparent reaction order of 0.5.

Earlier determinations on chars from high-rank coals indicated a value of n = 1. However, a reanalysis (8) of Field's data (2) for a low-rank coal char showed in this case that there is little to choose between n equal to 0.5 or 1.0.

Earlier work in this laboratory has shown that combustion of chars of low and high-rank coals occurs under regime II conditions (3), i.e. under circumstances where pore diffusion as well as chemical reaction exercises marked rate control. The values of the activation energy for the Millmerran and Wandoan chars (~ 20 kcal/mol), together with those of the other materials listed in Table II, correspond to the value expected for the combustion of impure carbons in the regime II region (3).

Confirmatory evidence that these materials are indeed burning under regime II conditions is provided by changes in the particle structure. In the present work, the particle size of the chars decreased and the particle density diminished slightly as the particles burned away, indicative of the expected changes for regime II conditions. This behavior is illustrated in Figure 5 by the data for chars produced from Millmerran and Wandoan coal at the lowest pyrolysis temperatures. The data lie mainly between the theoretical curves calculated for particles burning at constant density and decreasing size, or at constant size and decreasing density.

Conclusions

The reactivities of chars, at an oxygen pressure of 1 atm, produced from Millmerran subbituminous coal decrease with increasing pyrolysis temperature and are similar in magnitude to the reactivities of chars derived from a brown and a bituminous coal and to the reactivities of anthracite and semianthracite. However, the reactivity of Wandoan char, also of subbituminous origin, is about twice the reactivity of Millmerran char and about ten times the reactivity of petroleum coke. Taking into account the observed activation energy values, and the changes in particle size and particle density with burn-off, it is concluded that the combustion rates of Millmerran and Wandoan chars are controlled by the combined effects of pore diffusion and chemical reaction.

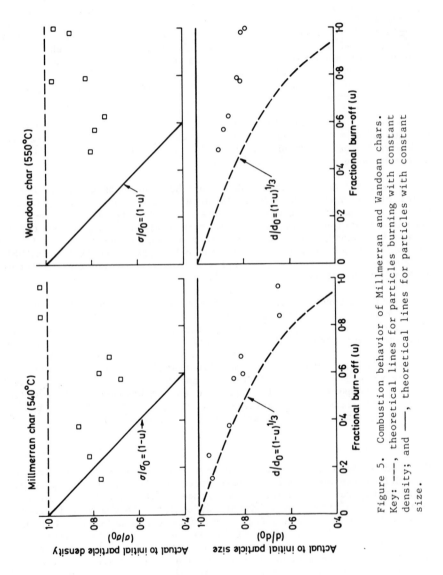

Figure 5. Combustion behavior of Millmerran and Wandoan chars. Key: ---, theoretical lines for particles burning with constant density; and ——, theoretical lines for particles with constant size.

Acknowledgments

The author is grateful to R.J. Hamor for carrying out the experimental measurements and to I.W. Smith for valuable discussions.

Literature Cited

1. Hamor, R.J.; Smith, I.W.; Tyler, R.J. Combust. Flame 1973, 21, 153-162.
2. Field, M.A. Combust. Flame 1969, 13, 237-252.
3. Mulcahy, M.F.R.; Smith, I.W. Rev. Pure Appl. Chem. 1969, 19, 81-108.
4. Mulcahy, M.F.R.; Smith, I.W. In "CHEMECA 70"; Butterworth: Australia,1971; pp. 101-118.
5. Smith, I.W. Combust. Flame 1971, 17, 303-314.
6. Smith, I.W. Combust. Flame 1971, 17, 421-428.
7. Sergeant, G.D.; Smith, I.W. Fuel 1973, 52, 52-57.
8. Young, B.C.; Smith, I.W. In "Eighteenth Symposium (International) on Combustion"; The Combustion Institute: Pittsburgh, 1981; pp. 1249-1255.
9. Edwards, J.H.; Smith, I.W.; Tyler, R.J. Fuel 1980, 59, 681-686.
10. Smith, I.W. In "Nineteenth Symposium (International) on Combustion"; The Combustion Institute: Pittsburgh, 1983; pp. 1045-1065.
11. Smith, I.W.; Tyler, R.J. Combust. Sci. Technol. 1974, 9, 87-94.
12. Field, M.A.; Gill, D.W.; Morgan, B.B.; Hawksley, P.G.W. "Combustion of Pulverized Coal"; BCURA: Leatherhead, 1967; pp. 186-210.
13. Ayling, A.B.; Smith, I.W. Combust. Flame 1972, 18, 173-184.
14. Young, B.C. In "Proceedings of the International Conference on Coal Science"; Verlag Glückauf: Essen, 1981; pp 260-265.
15. Hamor, R.J.; Young, B.C. In "Proceedings of the 1983 International Conference on Coal Science"; Pittsburgh, 1983; pp. 575-578.

RECEIVED March 5, 1984

Cationic Effects During Lignite Pyrolysis and Combustion

BRUCE A. MORGAN and ALAN W. SCARONI

Department of Materials Science and Engineering, Pennsylvania State University, University Park, PA 16802

A Texas lignite was acid-washed to remove its ion exchangeable cations then alkaline-earth metals (Ca, Mg) were back exchanged in various concentrations. The raw and modified lignites were pyrolyzed and combusted in an entrained-flow reactor. The effect of the cations on pyrolysis and combustion behavior was investigated. Their presence markedly reduced the evolution of volatile matter during pyrolysis. This is attributed to an increase in secondary char-forming reactions involving the ion exchangeable cations and volatile matter molecules, particularly tars. The rate of combustion of the lignite was enhanced by the presence of Ca in ion exchangeable form. The comparative behavior of a lignite treated with Ca, a good catalyst for the $C-O_2$ reaction, and Mg a poor catalyst, indicates that the heterogeneous $C-O_2$ reaction may be important during a short intermediate stage of lignite combustion.

In the U.S. a major usage of fuels is for electric power generation (1). Fuel oil is used to generate about 15% of the nation's electricity (1). However, it is possible with present technology along with new developments in coal combustion to reduce reliance on petroleum-based fuels by replacing them with coal. This would make more oil available to other areas such as the production of chemicals and pharmaceuticals.

The purpose of the present work is twofold. First, it is aimed at gaining understanding of the fundamentals of lignite combustion and the factors which most influence combustion behavior. This is predicated on the fact that the U.S. contains vast quantities of low-rank coals, including lignites. Recent estimates put these deposits at approximately 38 billion tons (2). An advantage of using lignites in industrial combustors and boilers is their generally low sulfur content. The formation of SO_2 and SO_3 within the combustor will be minimal, thereby reducing environmental concerns associated with acid rain.

0097-6156/84/0264-0255$06.00/0
© 1984 American Chemical Society

There can be problems associated with the use of lignites for power generation, however. In general, low-rank coal deposits are located in sparsely populated areas, for example, in the Northern Great Plains Province. Also, lignites have a propensity to slag in combustors and form deposits on heat transfer tubes in boilers. This has been attributed to the presence of alkali and alkaline-earth metals which exist, in part, as ion-exchangeable cations located on the carboxyl groups of low-rank coals (3). Removal of these cations may produce a more desirable fuel which could conceivably be transported to large population centers. As examples, coal-water and coal-liquid CO_2 slurry pipelines are currently being advocated as effective modes of transportation. The coal-liquid CO_2 pipeline is favored by some because of the unavailability of large quantities of water in some western states.

The second objective of this research is to modify, and thereby improve, the combustion behavior of lignites so that they can be used more extensively as an energy source. In particular, it is considered necessary to know whether or not the presence of ion-exchangeable cations significantly affects the combustion behavior of lignites. This would indicate whether there is any practical advantage of removing or adding additional cations prior to combustion.

American lignites have primarily alkali and alkaline-earth-metals exchanged on their carboxyl groups (4). Tanabe (5) suggests that these metal oxides act as polymerization catalysts for hydrocarbons. Longwell et al. (6) are currently investigating the effect of calcium oxide on the cracking of aromatics and other hydrocarbons. They find that calcium oxide cracks aromatics more efficiently than it does other hydrocarbons such as aliphatics. However, the finding that coke and tar were the principal products implies that calcium oxide was selectively polymerizing rather than cracking the aromatic compounds.

Pyrolysis occurs to some extent in all coal conversion processes. Much work has been done on the effect of various parameters such as atmosphere, temperature and heat flux on the pyrolysis of coal (7,8). The present work, however, is concerned more with the effect of an indigenous property of coal, its inorganic constituents, rather than processing variables. Morgan (4) has discussed in detail how mineral matter in low-rank coal exists in three distinct forms; as discrete minerals such as clays, as mineral matter associated with the organic matter such as ion-exchangeable cations and as trace elements. During combustion, the mineral forms undergo physical and/or chemical transformations leading to slag and deposit formation on various heat transfer surfaces in combustors and boilers. This has the effect of reducing the heat transfer efficiency of the boiler, and, to some extent, combustion efficiency because of drastically reduced free space in the combustor, thereby reducing particle residence times below those required for complete combustion of the fuels.

Several researchers (9,10) have studied the effect of ion-exchangeable cations on weight loss rate during rapid pyrolysis, that is pyrolysis at heating rates of 10^4-10^5 K/s. There is general consensus that ion exchangeable cations promote secondary char-forming reactions (cracking and/or polymerization), thereby reducing volatile

matter yields and changing product compositions. Specific mechanisms
for these reactions have not been determined. However, it is suggested that ion-exchangeable cations either react chemically with
the volatile matter or prevent the escape of volatile matter molecules from coal particles by physically blocking pores which act
as exit routes. Upon investigation of the volatile matter composition, Tyler and Schafer (11) found no large difference in the quantity and quality of light hydrocarbon gases evolved from acid-washed
coals. Thus it appears, as was suggested by Tanabe (5), that cations
selectively polymerize rather than crack aromatic tars.

Heterogeneous Combustion. An attendant difficulty in studying the
effects of ion-exchangeable cations on the heterogeneous char combustion stage of lignite combustion is the existence of a concurrent
or at least overlapping pyrolysis process. To overcome this difficulty, many researchers purposely separate the pyrolysis stage from
the char combustion step. Generally, the coal is first pyrolyzed
under controlled conditions then heterogeneous char combustion is
investigated subsequently. For example, this approach has been used
widely (12-14) to investigate and compare the effect of alkali and
alkaline-earth metals on the gasification behavior of various carbon
and coal chars. Walker et al. (12) and McKee (13) have done extensive work in a thermogravimetric analyzer at 0.1 MPa of air at various temperatures on the reactivity of graphite impregnated with various metals. They found that alkali and alkaline-earth metals, specifically Na and Ca, are excellent catalysts for the $C-O_2$ reaction.
This was attributed to the ability of the catalysts to undergo redox
reactions. That is, they believed Ca, for example, catalyzed the
reaction by supplying oxygen for the $C-O_2$ reaction according to the
following mechanism:

$$CaO + \tfrac{1}{2} O_2 \rightarrow CaO_2$$
$$CaO_2 + C \rightarrow CaO + CO$$

Radovic, Walker and Jenkins (14) studied the effect of various
forms of mineral matter in chars on their reactivity in air. It
was found that ion-exchangeable cations particularly Ca, greatly
enhanced char reactivity whereas discrete mineral phases had no significant effect. A redox reaction was again postulated as the mechanism of catalysis by the ion-exchangeable cations.
   As discussed previously, several attempts have been made to
interpret specific effects due to cations on the combustion behavior
of coals. However, most concentrate on studying the effect of cations on a single stage of the overall combustion process. Even
though this does improve understanding of the mechanisms involved
in each stage, it may not be appropriate when both pyrolysis and
combustion occur interactively. In coal combustion, these processes
overlap thereby affecting the chemistry occurring within each. This
project is an attempt to investigate, in situ, the effects of cations
on the individual and overall combustion processes.

Experimental

Lignite Preparation. The coal chosen is a Texas lignite, PSOC-623, from the Darco seam. The coal was obtained from the PSU-DOE coal sample bank and data base. This particular coal was chosen because it has been previously well characterized (4) and because it contains a large quantity of ion-exchangeable cations.

The as-received lignite had a large particle size distribution. To obtain a narrow particle size range, the coal was first crushed using a General Purpose Mill reducing all particles to below 1 mm in diameter. Next, the 1 mm particles were pulverized in a Hammer Screen Mill thereby reducing particle size to less than 150 μm.

The pulverized coal was sieved to separate a 200x270 mesh (U.S. Standard sieves) fraction. The mean weight particle size was 62 μm and the dispersion parameter 8.4 according to the Rosin-Rammler distribution (15). The size graded particles were dried under vacuum at 338 K for 8 hours, then stored in a dessicator until used.

Prior to combustion testing, 50 grams of the 200x270 mesh fraction were acid-washed in 900 ml of 0.4 M HCl. The coal and acid solution were stirred continuously for 24 hours. After that time, the coal was filtered and acid-washed a second time for 4 hours. The coal was again filtered and washed with 250 ml of cool, deionized water to remove chloride ions. Further washing to remove any remaining chloride ions was accomplished by stirring the coals with 900 ml of deionized water and heating to 238-258 K for 2 hours. This solution was filtered and checked for chloride ions by addition of a drop of silver nitrate to the filtrate. If reaction occurred, that is, if a cloudy white precipitate formed, the coal was washed repeatedly with cool deionized water until no reaction occurred upon the addition of silver nitrate. The acid-washed coal was dried under vacuum at 338 K for 8 hours.

The acid-washed sample was back exchanged with alkali (Na, K) and alkaline-earth (Ca, Mg) metals by placing 50 grams of the coal into 900 ml of 1 M metal acetate solution. The exchange time was varied according to the desired cation loading with a minimum time of 24 hours. The ion exchanged coal was washed with cool, deionized water (250 ml) and dried under vacuum at 338 K for 8 hours. Data for the lignite treated with Na and K will be reported elsewhere. Table I gives the ultimate analysis of the raw Texas lignite and the proximate analyses of the raw and modified coals. Acid-washing decreased the ash content of the coal by 33% (from 15.9 to 10.7 wt%). Subsequent addition of ion-exchangeable cations increased the ash content by between 18 and 30% depending on the loading. The volatile matter content (wt% daf) was lower for the acid-washed coal but was not significantly different for the cation-containing samples.

The distribution of the cations on the raw coal and the quantity of each cation back exchanged on the acid-washed coal is shown in Table II. These determinations were made by atomic absorption spectroscopy. The acid-washing was fairly efficient, removing over 99% of the cations present on the raw coal. Less than 0.01 wt% cations remained on the acid-washed lignite. The predominant cations on the raw coal were Ca and Mg, these two accounting for over 90% of the total quantity. As shown in Table II, the quantity of cations back exchanged on the acid-washed coal was greater than that

TABLE I

ULTIMATE ANALYSIS OF TEXAS LIGNITE PSOC 623 AND PROXIMATE ANALYSES OF RAW AND MODIFIED PSOC 623

	Ultimate					Proximate		
% (daf)	C	H	N	S	O (diff)			
	73.0	5.4	1.3	1.3	19.0			
						H$_2$O	Ash (dry)	VM (daf)
Raw						10.8	15.9	59.7
Acid Washed						7.8	10.7	54.9
Ca (1.0)						5.3	13.7	57.6
Ca (2.0)						9.2	13.7	59.1
Mg (1.3)						6.8	12.0	57.1

TABLE II

CATION LOADINGS ON RAW, ACID-WASHED AND BACK EXCHANGED PSOC 623

Raw	Ca	Mg	Na	K	Ba	Sr	Total	Acid Washed	Ca (1.0)	Ca (2.0)	Mg (1.3)
wt% (dry)	1.4	0.26	0.03	0.03	0.03	0.02	1.8	<0.01	1.9	4.1	1.6
meq/g (dry)	0.7	0.20	0.01	0.01	0.002	0.002	0.9	~0	1.0	2.0	1.3

contained on the raw coal on a meq/g basis. With the exchange procedure, it is difficult to obtain predetermined cation loadings.
However, a range of calcium loadings was obtained by varying the
exchange time. This allowed for evaluation of both the quantity
and type of cation on combustion behavior.

Selection of the cations was based on several criteria. First,
the raw coal contained alkaline-earth, mainly Ca and Mg, and some
alkali metals on its carboxyl groups. Also, McKee (13) and Walker
et al. (12) have shown that sodium, potassium and calcium are excellent catalysts for the $C-O_2$ reaction. Thus, these cations may have
a significant effect on the char burnout rate. In addition, Mg was
back exchanged on the coal since it was contained on the raw coal
and, as shown by Walker et al. (12), it is a poor catalyst for the
$C-O_2$ reaction. The purpose of using this alkaline-earth metal was
to determine if catalysis of the heterogeneous $C-O_2$ reaction affected
the char burnout rate. This would help to elucidate whether the
char burnout step was chemically or physically rate controlled.

Reactor System. The raw and modified coals were combusted in a hot-
wall, entrained-flow reactor at initial furnace (gas and wall) temperatures between 973-1173 K. Only data collected at 1173 K are presented here. The reactor is illustrated in Figure 1. It is essentially a vertical tube furnace which is heated externally by graphite
resistance heaters. Two separate gas streams are passed into the
reactor. The primary stream, He, is at room temperature and entrains
coal particles which enter the stream through an asymmetric venturi.
The entrained coal particles enter the reactor through a water-cooled
injector probe. The secondary stream, which can be either a reactive
or nonreactive gas, is preheated to a desired temperature then enters
the reactor through a flow straightener. The purpose of the flow
straightener is to assure a flat velocity profile at the tip of the
injector probe. The coal passes vertically down the center of the
reactor in a pencil stream and is collected by a water-cooled probe.
Particle residence times are changed by moving the collector probe
vertically inside the reactor and are calculated using a model formu-
lated by Morgan (9). Essentially, the model accounts for a slip
velocity between particles and flowing gas. Particle residence times
are a function of radial dispersion. The residence times reported
are averages for particles flowing down the central portion of the
reactor. The radius for collection equals half the radius of the
vertical reactor tube. Dispersion of the particles outside the collection radius was minimal. The relative error at a 95% confidence
level associated with particle residence times is 2%.

Chars are collected in a cyclone and tars in a filter after
being quenched in the collector probe. Weight loss data were determined using an ash tracer technique. There will be errors associated
with this technique if, for example, cation loss and/or sulfur fixation occurs to a different extent in the entrained-flow reactor than
during the ashing test. These errors are expected to be low for
the cations used in the present study and work is currently being
done to quantify them. The relative error at a 95% confidence level
associated with the weight loss determinations is a function of residence time. At short residence times, that is low levels of weight
loss, the relative error is up to 10%. At longer residence times,
hence higher weight losses, the relative error is less than 5%.

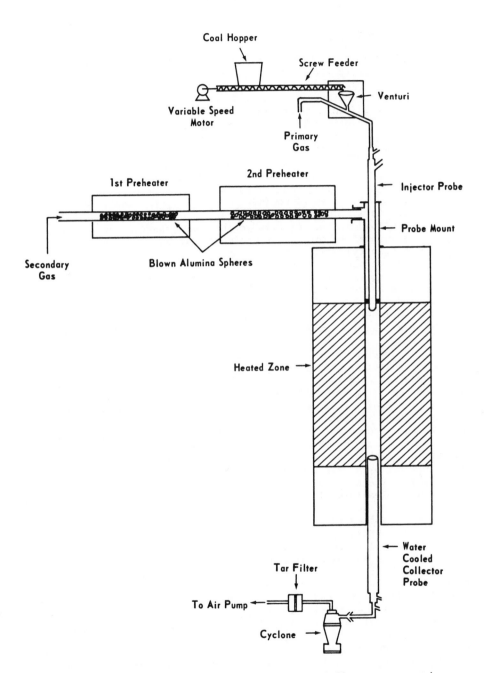

Figure 1. Schematic diagram of entrained-flow reactor and ancillary equipment.

Furthermore, the relative error was not affected by the presence
or absence of cations.

Results and Discussion

Pyrolysis Behavior. The effect of ion-exchangeable Ca on the pyrolysis of Texas lignite PSOC-623 is illustrated in Figure 2. This is
a plot of weight loss (daf basis) versus residence time for pyrolysis
in N_2 at a furnace temperature of 1173 K. Figure 2 illustrates several important points. First, ion-exchangeable calcium tends to
enhance the yield of volatiles at short residence times (<0.1 s).
The reason for this is not understood. However, the quantity of
volatile matter released from the coal at longer residence times,
above 0.1 s under these conditions, is reduced by the presence of
Ca. The acid-washed lignite which contained essentially no cations
lost more weight (50% daf) in the entrained-flow reactor than the
samples loaded with 1 meq/g Ca (42%) and 2 meq/g Ca (35%). Although
complete devolatilization of the samples does not occur in a residence time of 0.3 s, the relative weight loss behavior between the
samples is maintained for longer residence times. This has been
shown by Morgan (9), who captured samples in situ and held them at
the reaction temperature for 10 minutes. Therefore, the relative
weight loss behavior between the samples in the entrained-flow reactor is different from that displayed during the ASTM volatile matter
testing. In the latter case, the acid-washed coal lost less weight
than the cation-loaded samples. The reason for the different behavior under the different reaction conditions has not yet been established.

The reduction in volatile matter in the presence of Ca is attributed to an increase in secondary char-forming reactions (cracking
and/or polymerization). Based on a semi-quantitative analysis, 70%
by weight less tar was collected on the tar filter (see Figure 1)
for pyrolysis of the Ca loaded samples compared to the acid-washed
sample. Thus, it appears that ion-exchangeable Ca affects tar evolution either by chemically catalyzing reactions of tars or preventing
the escape of tars by physically blocking pores. A reduction in
tar yield was observed also by Morgan (9) and Franklin et al. (10)
under similar pyrolysis conditions.

Second, the data in Figure 2 indicate that, as the concentration
of ion-exchangeable Ca is increased from 1 to 2 meq/g, reduction
in the yield of volatile matter increases. That is, Ca (1.0) (1.0
denoting the concentration in meq/g of Ca), lost 7 wt% (daf coal)
more volatiles than Ca (2.0). This represents an enhancement in
the yield of volatiles of 20%. Work is continuing to determine precise mechanisms by which tar reduction occurs.

The relative effect of Ca and Mg on pyrolysis weight loss is
illustrated in Figure 3. Data are for a furnace temperature of
1173 K for Ca and Mg loadings of 1.0 and 1.3 meq/g, respectively.
For residence times below 0.15 s, Mg is more effective than Ca at
reducing the yield of volatile matter. This may be due to the 30%
higher Mg loading. In the total residence time of the reactor, however, Ca and Mg reduced the volatile matter yield equally. Tanabe
(5) has reported Ca to be a better polymerization and/or cracking
catalyst than Mg. The data in Figures 2 and 3 imply that in the

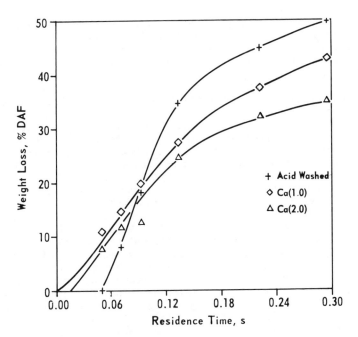

Figure 2. Pyrolysis weight loss for acid-washed and Ca loaded lignite, N_2, 1173 K.

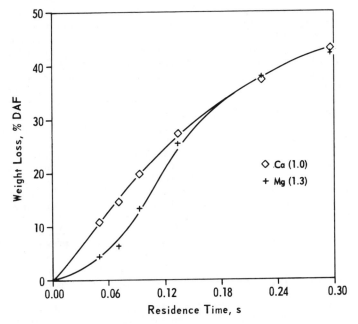

Figure 3. Pyrolysis weight loss for Ca and Mg loaded lignite, N_2, 1173 K.

very early stages of pyrolysis the presence of a good polymerization catalyst like Ca is not detrimental, rather is advantageous, to the yield of volatiles. At longer residence times, however, polymerization reactions become important and the presence of a good polymerization catalyst like Ca reduces the yield of volatiles.

Combustion Behavior. Figure 4 illustrates the effect of ion-exchangeable cations on weight loss of the lignite in air at an initial furnace temperature of 1173 K. Under the experimental conditions, the lignite was observed to ignite, hence, the weight loss was due to the combined effects of pyrolysis and heterogeneous char combustion. Figure 4 shows that the raw and acid-washed lignites had similar weight loss curves for residence times up to about 0.1 s. Also, above 0.12 s, weight loss rates were similar. In the region between 0.1 and 0.12 s, however, the acid-washed lignite lost less weight than the raw lignite. This may be due to catalysis of heterogeneous char combustion by Ca cations during this brief period. This trend is relevant to proposed mechanisms of lignite combustion.

Figure 5 illustrates the relative behavior of a Ca loaded sample, Ca being a good catalyst for the $C-O_2$ reaction and the acid-washed lignite. In the early stages of combustion (< 0.1 s), both the acid-washed and Ca loaded samples had similar weight losses with weight loss from the Ca loaded sample being only slightly larger than for the acid-washed sample. This implies that pyrolysis is a dominant mechanism at short residence times.

Again, however, it appears that there is an important intermediate stage between pyrolysis and heterogeneous char combustion. This occurs at residence times between 0.1-0.12 s for the current experimental conditions. As illustrated in Figure 5, Ca loaded samples lose weight more rapidly than acid-washed samples during this intermediate stage. This may be attributed to the catalysis of the $C-O_2$ reaction. At longer residence times (> 0.12 s), however, weight loss rates are similar for the Ca loaded and the acid-washed samples indicating possibly a physical controlling mechanism for char burnout.

To elucidate further the initial stage of combustion, the relative behavior of Mg loaded and acid-washed samples was compared. Walker et al. (12) have shown that magnesium is a poor catalyst for the $C-O_2$ reaction. Therefore, the addition of magnesium to the acid-washed coal should not increase weight loss in the important intermediate stage by catalyzing significantly the $C-O_2$ reaction. If this is the case, then the relative behavior in N_2 and air should be similar. Figure 5 illustrates that there is an effect on weight loss by the addition of Mg. For residence times < 0.1 s, the Mg loaded sample lost less weight than the acid-washed sample. At longer residence times, the weight loss by both the acid-washed lignite and the Mg loaded samples was very similar. This parallels the general behavior during pyrolysis in N_2 and implies no significant catalysis of the $C-O_2$ reactions at any stage of combustion.

In addition, for similar loadings of Ca and Mg, 1.0 and 1.3 meq/g, respectively, the Ca loaded sample lost more weight than the Mg loaded sample for all residence times. This implies that for combustion, Ca has a dual advantage over Mg. First, at short residence times when pyrolysis dominates, volatile matter yield is higher. Second, at intermediate residence times, catalysis of the $C-O_2$ reaction occurs. There is probably no advantage in the final char burnout stage, however.

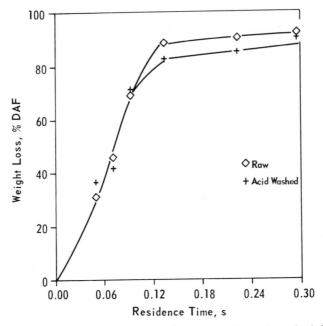

Figure 4. Combustion weight loss for raw and acid-washed lignite, air, 1173 K.

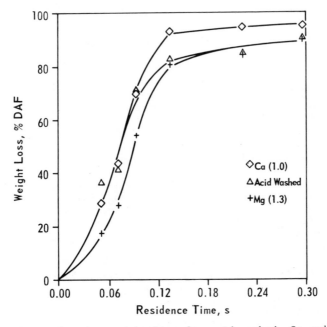

Figure 5.   Combustion weight loss for acid-washed, Ca and Mg loaded lignite, air, 1173 K.

Conclusions

It has been suggested previously that low rank coal combustion occurs in two distinct stages: pyrolysis and heterogeneous char combustion. In the past, in order to study the effect of various parameters on coal combustion, these stages were purposely separated. The effect of cations on pyrolysis, for example, has been studied under rapid heating conditions such as those obtained in an entrained-flow reactor. In addition, cationic effects on heterogeneous char combustion have been investigated in TGAs. This paper has shown that this experimental approach may not be applicable to combustion of low rank coals in industrial furnaces. That is, there appears to be three important regions for lignite combustion. The first region is controlled by pyrolysis and can be modeled by pyrolysis in N_2 under similar heating rates. In a second short but important region, heterogenous char combustion dominates and catalysis of the $C-O_2$ reaction is important. It appears that, in this region, chemical reactivity controls or at least affects the rate of weight loss. Although different conditions prevail, chemical reactivity data obtained from chars in TGAs may indicate qualitative behavior in the region. The rate of weight loss in the third region appears to be independent of the presence of cations. This implies a physically rate controlled mechanism. Therefore, chemical reactivity data for chars would not correctly model this final burnout stage.

Acknowledgments

The lignite was supplied by W. Spackman from the PSU-DOE Coal Sample Bank and Data Base. Both The Gulf Oil Foundation and the PSU Coal Cooperative Program provided financial support for Bruce Morgan.

Literature Cited

1. Gulf Oil Corporation "Some Useful Facts on Energy," January 1981.
2. American Coal Foundation "The Modern Energy Miracle," 1983.
3. Baria, D. N., Kube, W. R. and Paulson, L. E. Proc. of the 64th CIC Coal Symposium, 1982, p. 288.
4. Morgan, M. E.; Jenkins, R. G. and Walker, P. L., Jr. Fuel, 1981, 60, 189.
5. Tanabe, K. "Solid Acids and Bases," Acad. Press, NY, 1970.
6. Longwell, J. P.; Chiu, K. L.; Williams, G. C. and Peters, W. A. DOE Progress Report, Contract #DE-FG22-80PC-30229, September 1982 and June 1983.
7. Howard, J. B. In "Chem. of Coal Utilization," Elliot, M. A., Ed. Wiley-Interscience: NY, 2nd Suppl. Vol., 1981, p. 665.
8. Freihaut, J. D.; Zabielski, M. F.; Seery, D. J. Nineteenth Symp. (Int.) on Comb., The Comb. Institute, 1982, p. 1159.
9. Morgan, M. E. Ph.D. Thesis, The Pennsylvania State University, 1983.
10. Franklin H. D.; Cosway, R. G.; Peters, W. A. and Howard J. B. Ind. Eng. Chem. Process Des. Dev. 1983, 22, 39.
11. Tyler, R. J. and Schafer, H. N. S. Fuel 1980, 59, 487.
12. Walker, P. L., Jr.; Matsumoto, S.; Hanzawa, T.; Muira, T. and Ismail, I. Fuel 1983, 62, 170.
13. McKee, D. W. Fuel 1983, 62, 170.
14. Radovic, L. R.; Walker, P. L. Jr. and Jenkins, R. G Fuel 1983, 62, 209.
15. Rosin, P. and Rammler, E. J. Inst. Fuel 1933, 7, 29.

RECEIVED March 15, 1984

Catalysis of Lignite Char Gasification by Exchangeable Calcium and Magnesium

T. D. HENGEL[1] and P. L. WALKER, JR.

Department of Materials Science and Engineering, Pennsylvania State University, University Park, PA 16802

A Montana lignite was treated with HCl and HF to remove essentially all inorganic constituents and place all carboxyl groups in the proton form. The treated lignite was then exposed to Ca and Mg acetate solutions, either separately or jointly, to add controlled amounts of cations to the carboxyl groups. Following pyrolysis of the lignites, reactivities of their chars in air, CO_2, and steam were measured. The presence of Mg is seen to not affect the excellent catalytic activity of Ca for char gasification.

The high intrinsic reactivity of chars derived from lignite coals is primarily due to the presence of well dispersed metal oxide particles, which are formed when carboxyl groups on the lignite surface, having been exchanged with various cations from the local groundwater, decompose as the lignite is heated to gasification temperatures. Roughly half of the carboxyl groups on American lignites have undergone exchange, with Ca and Mg predominantly, and to lesser extents with other alkali and alkaline-earth cations (1). These cations have been shown to be good catalysts for lignite char gasification in air, CO_2 and steam (2).

It is noted, however, that lignite char reactivity decreases rapidly as final heat treatment temperature increases (3,4) or as soak time at temperature increases (4). Radovic and co-workers (5) have shown that the majority of this decrease can be correlated with a decrease in CaO dispersion in the case of Ca loaded demineralized (Dem) lignite chars.

Exchangeable cations play an important role in the behavior of lignite coals in coal conversion processes. It is, therefore, important that a fundamental understanding be attained, first of the possible interaction of exchangeable cations in lignites with each other, and, second, of the possible effects on the subsequent catalytic activity of the cations for lignite char gasification.

[1] Deceased

0097-6156/84/0264-0267$06.00/0
© 1984 American Chemical Society

The aim of this study was to develop such an understanding using the two most abundant cations in American lignites: Ca and Mg. The objectives of this work was to study the effect which adding Mg, which is a poor gasification catalyst, to Ca, which is a good gasification catalyst, has on the catalytic behavior of Ca for lignite char gasification. That is, will the presence of Mg in some way affect the sintering of Ca and, hence, affect the catalytic activity of Ca?

Experimental

A Montana lignite (6) was subjected to essentially complete demineralization with HCl and HF in order to remove essentially all the inorganic constituents (exchangeable cations and mineral matter) present (1). The Dem lignite was then ion-exchanged with Ca or Mg using solutions of their acetate salts. Seven levels of Ca and three levels of Mg were obtained using loading solutions of varying concentrations. A sample of Dem lignite was also co-exchanged with Ca and Mg in order to study the effect of their interaction on subsequent char reactivity. Chars were prepared by slowly heating (10 K/min) the treated lignites in a box furnace equipped with an air tight retort under N_2 to final temperatures of 973 or 1273 K and held (soaked) for 1 h. Char reactivities were determined by isothermic thermogravimetric analysis (TGA) in 0.1 MPa air, 0.1 MPa CO_2 and 3.1 kPa steam (saturated N_2). The maximum slope (R_{max}) of the TGA recorder plot was used as a measure of gasification reactivity (normalized to initial weight of char, daf basis). From a series of preliminary runs (7), reaction conditions were selected (4 mg of char spread thinly on a Pt pan and sufficiently low temperatures) in order to eliminate heat and mass transfer limitations. Therefore, reported reactivities are believed to be intrinsic, chemically controlled rates.

   Selected chars were examined by x-ray diffraction (XRD) to identify Ca and Mg containing species present in the chars. Average crystallite diameters were determined using the line broadening concept and the Scherrer equation (8).

Results and Discussion

Figures 1 and 2 show the reactivity results in 0.1 MPa air for the 973 and 1273 K chars, respectively. As seen previously, Ca is a good catalyst for char gasification in air, while Mg is a poor char gasification catalyst (2). For Ca loaded chars at both heat treatment temperatures, it is seen that two distinct regions of reactivity behavior are present. For loadings up to ~4 wt% Ca, the reactivity increases linearly with increasing Ca content. For loadings greater than ~4 wt% Ca, however, reactivity remains essentially constant with increasing Ca content. This is contrary to the behavior seen by Hippo et al. (9), who reported a linear increase in reactivity with increasing Ca content up to 12.9 wt% in 0.1 MPa steam for a Texas lignite. Since mass transport limitations are believed to be absent, the plateau is not due to diffusional effects. This suggests that the plateau is the result of a catalyst "saturation effect," that is, adding of catalyst beyond a certain amount does not further increase the observed rate. For the relatively small range

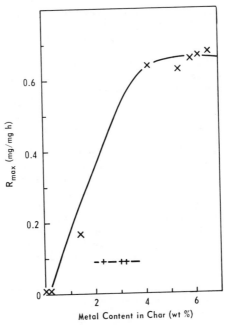

Figure 1. Reactivities of 973 K chars in 0.1 MPa air at 603 K.
Key: X, Dem + Ca; and +, Dem + Mg.

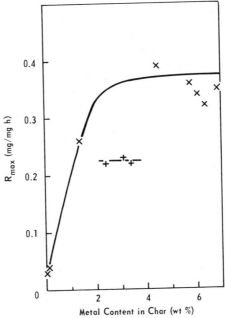

Figure 2. Reactivities of 1273 K chars in 0.1 MPa air at 603 K.
Key is the same as in Figure 1.

of Mg contents a similar saturation effect appears to be operating. Examination of the cation exchanged 1273 K chars by XRD revealed that CaO and MgO were the major species present in the Dem+Ca and Dem+Mg chars, respectively.

Reactivity results for the 1273 K chars reacted in 0.1 MPa CO_2 and 3.1 kPa steam at 1053 K are shown in Figures 3 and 4, respectively. The same trends evident for the 1273 K chars reacted in 0.1 MPa air are also seen in these reactant gases. Two regions of reactivity behavior are present in both cases. Calcium is seen to be a good char gasification catalyst, while Mg has very little catalytic effect.

The purpose of this work was to study the effect that adding Mg to Ca has on the subsequent catalytic behavior of Ca for lignite char gasification. To study this effect, the chars of the co-exchanged Dem lignite were used. It was reasoned that the co-exchanged chars should have a Ca content which was in the region where reactivity increased with metal content, as to avoid the saturation region. Also, the char should contain more Mg than Ca on a molar basis, since an effect is more likely to be observed under these conditions. Metal contents of the co-exchanged chars are presented in Table I.

Table I. Metal Contents in the Co-Exchanged Chars (Dry Basis)

Char	Calcium		Magnesium	
	mmol/g	wt%	mmol/g	wt%
973 K	0.54	2.2	0.91	2.2
1273 K	0.56	2.2	0.94	2.3

To determine if there was any effect, reactivity data of the mono-exchanged chars were used to calculate an expected reactivity. Calculation of these expected reactivities was straight forward, which assumed that the catalytic activities of Ca and Mg were simply additive. From the metal contents in Table I, the contributions of Ca and Mg to the calculated reactivities could be determined from Figures 1-4 by reading the reactivity value corresponding to that amount of Ca and Mg directly from the appropriate figure. The reactivity of the Dem-char is subtracted once because it is included in both the Ca and Mg contributions, while it is included once in the co-exchanged char. This step is not necessary when the rate of the uncatalyzed (Dem) reaction is low; but is necessary when the uncatalyzed rate is high, as is the case of CO_2 (Figure 3). For consistency, the Dem reactivity was always subtracted once. The calculated reactivity is the sum of the Ca and Mg contributions, with the rate of the Dem reactivity subtracted once. By comparing the calculated reactivity with that actually observed for the co-exchanged chars, one will be able to tell if indeed there is an effect. The calculated and observed reactivities are given in Table II for the various gases.

It is quite clear from Table II that the addition of Mg to Ca does not have a significant effect on the subsequent activity of Ca. The observed reactivities are, within experimental error, the same as the calculated reactivities. This shows that the catalytic

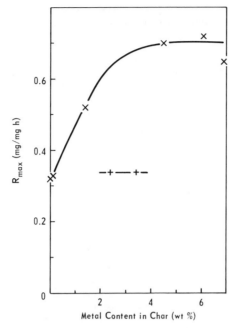

Figure 3.   Reactivities of 1273 K chars in 0.1 MPa CO_2 at 1053 K.
Key is the same as in Figure 1.

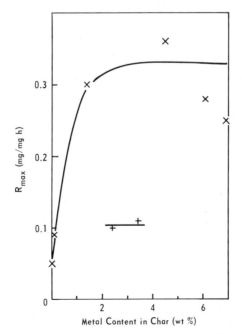

Figure 4.   Reactivities of 1273 K chars in 3.1 kPa steam at 1053 K.
Key is the same as in Figure 1.

Table II. Calculated and Observed Reactivities of the Co-Exchanged
           Chars

Char	Reactivity Contributions (mg/mg h)			Reactivities (mg/mg h)	
	Ca	Mg	Dem	Calculated	Observed
			Air		
973 K	0.31	0.09	0.01	0.39	0.42
1273 K	0.34	0.22	0.03	0.53	0.60
			CO_2		
1273 K	0.60	0.34	0.32	0.62	0.64
			Steam		
1273 K	0.30	0.10	0.05	0.35	0.31

effects of Ca and Mg are additive and they do not interact in a significant way to either increase or decrease their combined catalytic
effects. This is perhaps due to the fact that MgO is not significantly soluble in CaO below 1900 K (10). The presence of MgO does
not reduce the sintering of CaO. This is supported by examination
of the 1273 K co-exchanged char by XRD. The average crystallite
diameter of the CaO was calculated to be 30 nm. This is the same
value obtained for a 1273 K Dem+Ca char that has a similar Ca content (1.4 wt%).

Conclusions

The addition of Mg to Ca does not have a significant effect on the
subsequent activity of Ca for lignite char gasification under the
slow heating conditions used in this study. Their catalytic effects
were found to be additive, indicating that they did not interact
in any significant way. This was supported by XRD results which
showed that the sintering of CaO was not reduced by the presence
of MgO.

Acknowledgments

This research was supported, in part, by Contract No. 956112 from
the Jet Propulsion Laboratory. Professor W. Spackman kindly supplied
the coal sample studied.

Literature Cited

1. Morgan, M. E.; Jenkins, R. G.; Walker, P. L., Jr. Fuel 1981,
    60, 189.
2. Walker, P. L., Jr.; Matsumoto, S.; Hanzawa, T.; Miura, T.;
    Ismail, I. M. K. Fuel 1983, 62, 140.

3. Jenkins, R. G.; Nandi, S. P.; Walker, P. L., Jr. Fuel 1973, 52, 288.
4. Radovic, L. R.; Walker, P. L., Jr.; Jenkins, R. G. Fuel 1983, 62, 209.
5. Radovic, L. R.; Walker, P. L., Jr.; Jenkins, R. G. J. Catalysis 1983, 82, 382.
6. Penn State/DOE Coal Data Base, PSOC 833, Coal Research Section, The Pennsylvania State University.
7. Hengel, T. D. M.S. Thesis, The Pennsylvania State University, 1983.
8. Klug, H. P.; Alexander, L. E. "X-ray Diffraction Procedures"; Wiley, New York; 1974, p. 491.
9. Hippo, E. J.; Jenkins, R. G.; Walker, P. L., Jr. Fuel 1979, 58, 338.
10. Doman, R. C.; Barr, J. B.; NcNally, R. N.; Alper, A. M. J. Amer. Cer. Soc. 1963, 46, 313.

RECEIVED February 17, 1984

Mechanistic Studies on the Hydroliquefaction of Victorian Brown Coal

F. P. LARKINS[1], W. R. JACKSON, D. RASH, P. A. HERTAN, P. J. CASSIDY, M. MARSHALL, and I. D. WATKINS

Department of Chemistry, Monash University, Clayton, Victoria 3168, Australia

The overall aim of our recent studies has been to obtain a more complete understanding of the mechanisms for the principal reactions which occur during the catalyzed hydroliquefaction of low-rank, high oxygen containing (ca. 25 wt% db) coals. The results of 70 ml batch autoclave studies with and without added catalysts on Victorian brown coal, on a number of different coal-derived products and on related model ether compounds are discussed herein and in previous publications (1-6). On the basis of these investigations a mechanism is proposed for the hydroliquefaction process which emphasizes the role of catalysts inhibiting repolymerisation reactions, the significance of interconvertibility of coal-derived products and the importance of hydrogen donation from molecular hydrogen and the vehicle tetralin.

EXPERIMENTAL

Coals: Two samples of dried (105°C, under N_2) Victorian brown coals from the Morwell seam were used in these experiments, viz., Drum 66 (1976 100t bulk sample) and Drum 289 (1979 100t bulk sample). The chemical characteristics of the coals on a weight per cent dmmf basis are as follows: Drum 66, C, 69.3; H, 5.0; O (by diff.) 24.5; N, 0.6; S, 0.6. Acidic O, 9.7 mol kg^{-1} dmmf coal, non-acidic O, 5.6 mol kg^{-1} dmmf coal (2). Drum 289, C, 69.2; H, 4.8; O (by diff.) 25.2; N, 0.52; S, 0.25. Acidic O, 9.6 mol kg^{-1} dmmf coal; non-acidic O, 5.8 mol kg^{-1} dmmf coal. The elemental composition is therefore similar to that of a North Dakota lignite.

[1]Current address: Department of Chemistry, the University of Tasmania, Hobart, Tasmania 7001, Australia

0097-6156/84/0264-0275$06.00/0
© 1984 American Chemical Society

The coals (<250 μm) were hydrogenated in a 70 ml rocking autoclave
using a 1:1 slurry of tetralin:coal at initial hydrogen pressures of
1-10 MPa for 1 hour at a reaction temperature of 285°C. In related
studies temperatures in the range 345-460°C were used at 6 MPa
pressure. Both iron and tin-based catalyst systems were examined.

For the work reported herein the catalysts were incorporated by the
ion exchange technique (1). In other studies we have shown that
impregnation using water soluble metal salts is as effective in
achieving a comparable level of catalytic activity.

The products from these hydrogenations were separated into gases
(analyzed by G.C.), water (analysed by azeotropic distillation),
insolubles (CH_2Cl_2 insolubles), asphaltene (CH_2Cl_2 soluble/X4
insoluble) (Shell X4: 40-60°C b.p. light petroleum), oils (CH_2Cl_2
soluble/X4 soluble). Hydrogen transferred from the donor solvent
was determined by G.L.C. analysis of the ratio of tetralin to
naphthalene in the total hydrocarbon liquid product.

Coal derived materials: These products were obtained from our 1 kg
h^{-1} continuous reactor unit (7) as oils (X4 soluble) asphaltenes
(tetralin soluble/X4 insoluble) preasphaltene (also known as
asphaltol) (tetralin insolubles/tetrahydrofuran (THF) solubles) and
THF insoluble materials for subsequent reactivity studies.
Methylene chloride was not used in the separation because of concern
for its potential catalyst deactivation role in the subsequent series
of experiments to be undertaken. Experimental continuous reactor
conditions were a 3:1 slurry of tetralin:coal to which an iron (278 ±
30 mmol kg^{-1} dry coal from $FeSO_4.7H_2O$) and tin (20 mmol kg^{-1} dry
coal from $SnCl_2.2H_2O$) catalyst had been added. The reaction temperature was 400°C with a developed pressure of 10 MPa and a residence
time from 15 to 40 minutes.

Samples of each of the coal derived materials were reacted separately
in the presence of several catalysts in a 70 ml batch autoclave using
a 1:1 slurry of tetralin:material at 425°C with an initial hydrogen
pressure of 6 MPa for 1 hour at reaction temperature. The products
from these reactions were separated into oils, asphaltenes,
preasphaltene and THF insolubles.

Model ether systems: A series of ethers have been examined to assist
in the elucidation of the role of the iron and tin catalyst systems
investigated. Approximately one-third of the organically bound
oxygen in Victorian brown coal is in the form of ether linkages (8).
The ether linkages are up to 28 kJ mol^{-1} weaker than the equivalent
carbon-carbon linkages (e.g., $PhCH_2-OPh$ 221 kJ mol^{-1} cf. $PhCH_2-CH_2Ph$
249 kJ mol^{-1}) and thus provide a potential region for bond cleavage
within the brown coal structure which may be of central importance in
the overall liquefaction reaction sequence.

Hydrogenation conditions were 1-2 g model ether, initial hydrogen
pressure 6 MPa, reaction time 1 hour with catalyst loadings similar
to those used for coal hydroliquefaction studies. Further details
are outlined elsewhere (2,5).

RESULTS

Reactions with coal

Total conversion yields of the temperature range 345–460°C at 6 MPa hydrogen are summarized in Figure 1. Absolute conversions are lower than for work with 1 liter autoclaves and the continuous reactor unit because of different reactor temperature profiles and residence time; however, the qualitative conclusions are consistent. The effect of increasing the reaction temperature was to increase the total conversion of all reactions irrespective of the catalyst system.

The iron–tin catalyst system was the most temperature sensitive of the four coal-catalyst systems studied. Unlike the iron- and tin-treated coals which were only sensitive to temperature in the range 365–405°C, the iron–tin-treated coal was also sensitive to temperature in the range 385°–425°C where the conversion rapidly increased from 66% to 85% (daf coal). Thus the iron–tin system was most efficient at the higher reaction temperatures up to 425°C while the iron- and tin-treated coals were most effective up to 405°C (see Table 1, reference 2).

The initial hydrogen pressure dependent conversion results in the range 1–10 MPa are given in Figure 2 for studies at 385°C. The tin catalyzed reactions show most pressure dependence, consistent with previous findings that tin facilitates molecular hydrogen transfer during hydrogenation (4). For the untreated coal and for other catalyst systems there is only a small pressure dependence of product yields beyond 4 MPa.

A detailed study of the chemical constitution of the products revealed that their composition is influenced by temperature but not by pressure or the nature of the catalyst. With increased reaction temperature there was a decrease in total and acidic oxygen concentrations in the asphaltenes and a corresponding increase in both aromatic content and the C/H ratio. This observation is consistent with the loss of aliphatic side chains from the polycondensed ring systems.

Reactions with coal derived products

Hydrogenation studies were undertaken on the parent iron–tin treated coal (Drum 289) as well as the THF insolubles, preasphaltene, asphaltene and oil derived from a continuous reactor run as previously discussed. Studies with no additional catalyst added (case A) and with the addition of a sulphided nickel molybdate catalyst supported on alumina (case B) were performed. The results are presented in Table 1. The Ni/Mo catalyst in case B did not increase the conversion of the coal or the THF insolubles beyond that for case A because sufficient amounts of iron and tin materials were already

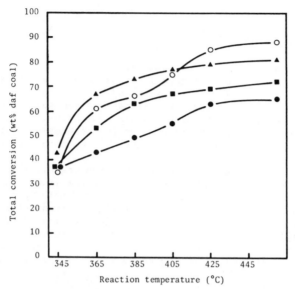

Figure 1. Total conversion versus reaction temperature.
Reaction conditions: 6 MPa initial hydrogen pressure;
1 h at temperature; 1:1 tetralin:dry coal;
●, Untreated coal; ■, iron-treated coal;
▲, tin-treated coal; ○, iron-tin-treated coal.

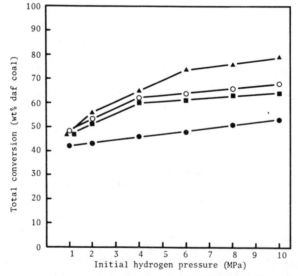

Figure 2. Total conversion versus initial hydrogen pressure.
Reaction conditions: 385°C reaction temperature;
1 h at temperature; 1:1 tetralin:dry coal;
●, Untreated coal; ■, iron-treated coal;
▲, tin-treated coal; ○, iron-tin-treated coal.

Table 1:   Product distribution from the catalyzed and uncatalyzed
reactions of Victorian Morwell coal and coal derived
reactants.

Reactant		P R O D U C T D I S T R I B U T I O N (wt.% reactant)			
		Oil	Asphaltene	Preasphaltene	THF Insoluble
Untreated coal[a]	A	26	7	1[c]	36[c]
Iron-tin treated coal[a]	A	42	14	1	17
Iron-tin treated coal[a]	B	40	10	2	20
THF[a] insolubles	A	39	13	2	25
	B	39	14	3	26
Preasphaltene[a]	A	44	28	5	13
	B	63	19	3	5
Asphaltene[a]	A	37	41	7	6
	B	67	22	2	2
Oil[b]	A	96	4	tr.	1
	B	97	tr.	1	3

A   No catalyst added.   B   Sulphided Ni/Mo catalyst (10% by weight reactant)

[a]   Reactions at 425°C for 1 h using 3 g reactant and 3 g tetralin with initial hydrogen pressure of 6 MPa.

[b]   Reactions as for a except 1 g oil and 5 g tetralin.

[c]   Estimated values (preasphaltene + THF insoluble = 37).

associated with these reactants to catalyze the reaction. The iron-tin treated coal has of course a much greater reactivity than untreated coal as shown in Table 1.   It is noteworthy that once isolated the THF insolubles showed a similar reactivity (conversion ∿75 wt% daf) to the coal (conversion ∿80 wt% daf).

The results in Table 1 show that for reactions at 425°C significant conversion of the preasphaltenes and the asphaltenes produced at 400°C to other products was possible.   In particular for the preasphaltenes >95% interconversion occurred, while for the asphaltene the interconversion was >59%.   A complete range of products was formed from high oil yields to repolymerized THF insoluble material.   This reactivity underlines the inherent instability of these intermediate products.   The addition of a sulphided Ni/Mo catalyst led to ∿50% improvement in oil yields,

however addition of the traditional first stage catalysts iron, tin
or the iron-tin mixture did not significantly improve the conversion
of asphaltene or preasphaltene to oil as compared to the reactions
without additive (results not presented here).   Tin, and to a lesser
extent, iron were successful in reducing the amount of repolymeriza-
tion of preasphaltene to the THF insolubles.

Hydrogenation of the oil fraction resulted in the formation of a
small amount of the higher molecular weight products (<4 wt%), but
the recovered oil contained a higher proportion of lower boiling
point material.

Reactions with model ether compounds

The reactions of a series of lignin-related model ether compounds
with iron(II) acetate and tin metal have been investigated (2,5).
The models selected contain phenoxy groups (PhO-$(CH_2)_n$Ph n = 1,2,3),
benzyl groups (PhCH$_2$-OCH$_2$Ph) and alkoxy groups (PhCH$_2$CH$_2$-OCH$_2$CH$_2$Ph).
Conversion results for the reactions of 2-phenyl ethyl phenyl ether
(PhO-CH$_2$CH$_2$Ph) and dibenzyl ether (PhCH$_2$OCH$_2$Ph) are shown in Figure
3.   For the 2-phenylethyl phenyl ether the extent of conversion is
significantly reduced in the presence of both metal additives with
iron being more effective than tin in suppressing the decomposition.
For example at 325°C there was only 13% conversion in the presence
of an iron based additive compared with 33% with tin and 43% without
additive.   In contrast, addition of an iron-based catalyst to a
reaction of dibenzyl ether (Figure 3B) was found to promote the
conversion while the tin-based catalyst suppressed it relative to no
additive being present.   At 300°C the conversion was 96% with iron,
12% with tin and 19% without additive.   The thermal decomposition of
these ethers is believed to be via a radical chain mechanism (10).
It is postulated that while iron catalysts may facilitate carbon-
oxygen bond cleavage, as evidenced by the increased reactivity of
benzyl and aliphatic ethers, when phenoxy radicals are produced they
are strongly adsorbed on the catalyst surface.   They are not therefore readily available to propagate the radical chain reactions.
This effect is restricted to the phenoxy radical and not observed for
a species such as PhCH$_2$CH$_2$O· in the reaction of PhCH$_2$CH$_2$OCH$_2$CH$_2$Ph
because the resonance stabilization energy associated with the PhO·
would result in it having a longer half-life and hence time to
diffuse to the catalyst surface without being stabilized by hydrogen.
The iron catalyst does not show the same affinity for the benzylic
radical even though it is also a long-lived species.

The tin additive is present in the liquid state under the conditions
of the present experiments.   It has a smaller inhibiting effect than
iron on the reactivity of the phenoxy and benzyl ethers.   Two
explanations are plausible.   Hydrogen dissolved in the tin may react
with the benzyl and phenoxy radicals which are the chain propagators
and remove them from the system.   The rate of bond cleavage is
therefore lowered.   Alternatively, tin may promote radical
recombination reactions.   By either route the tin would be acting
to inhibit propagation reactions.

A. 2-Phenylethyl phenyl ether   B. Dibenzyl ether

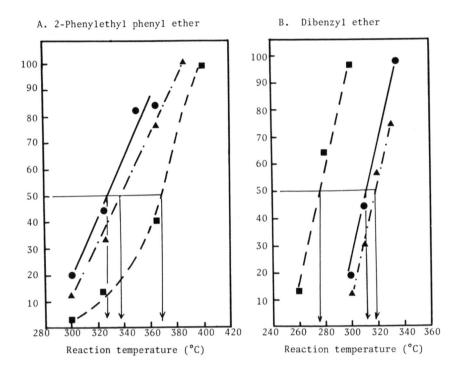

Figure 3.   The effect of iron and tin on the conversion of model
            ethers.   Reaction conditions: 6 MPa initial hydrogen
            pressure; 1 h at temperature.   ●, Untreated;
            ■, iron-treated; ▲, tin-treated.

Role of Radical Propagation Reactions

If the iron- and tin-based catalyst indeed acts to inhibit radical
propagation reactions then two major consequences of this effect can
be tested in coal reactions. Firstly, and most importantly, the
rate of the thermally induced breakdown of the coal should be reduced
in the initial minutes of the coal reaction. This has recently been
demonstrated when iron catalysts were used in a series of <u>short</u>
residence time studies (∿12 minutes to temperature then autoclave
quenched). The results, summarized in Table 2, show that the
conversion of acid washed coal after rapid heating to 450°C had a
conversion of 55% (daf coal) while the conversion of acid washed coal
that had been re-exchanged with 240 mmol of iron kg^{-1} (dry coal) had
a conversion of only 46%. A similar trend was found for the reaction
of "as mined" coal (39% conversion) and the same coal that had been
treated with a solution of iron(II) acetate (300 mmol kg^{-1} dry coal)
and had a conversion of 33%. It is interesting to note that the
natural inorganics of the "as-mined" coal had a very strong inherent
activity to slow the rate of thermal decomposition.

Table 2:   The effects of iron in short residence time
            hydroliquefaction reactions of Victorian (Morwell)
            brown coal

COAL TREATMENT	ACID WASHED	ACID WASHED +Fe	UNTREATED COAL	UNTREATED COAL +Fe
conversion	55	46	39	33

conversion = 100 - % THF insolubles

The second effect of the reduction in the rate of radical propagation
reactions is to slow down the rate of radical recombination or
polymerization.   This effect has been demonstrated in hydrogenation
reactions of preasphaltene in the presence of the non-hydrogen donor
solvent, decalin.   The results are summarized in Table 3.

Table 3:   The effect of iron and tin on the product distribution
            for preasphaltene hydrogenation using decalin as a
            solvent.

	Product Distribution (wt% initial preasphaltene)		
catalyst	none	iron	tin
oil	25	28	30
asphaltene	16	20	33
preasphaltene	8	10	10
THF insolubles	40	30	15

In the reaction of untreated preasphaltenes the yield of THF insoluble "residue" was 40% by weight of the reactant asphaltol while the
presence of iron reduced the residue yield to 30%.   Tin had the
largest effect in these reactions reducing the residue yield to 15%.
The inter-conversions from reactions of the untreated, iron- and tin-
treated preasphaltenes were all 91 ± 1% and so the reduced polymerization resulted in increased yields of oil and asphaltene.

MECHANISTIC CONSIDERATIONS

Many reaction schemes have been proposed to interpret the results of
hydroliquefaction experiments following the pioneering work of Weller
et al. in 1951 (9).   Factors to be taken into account when considering the complex liquefaction process include

    1.   Coal-solvent interactions leading to swelling and
          dissolution of the coal (11, 12).

2. The radical production resulting from thermal degradation of coal molecules (13,14).

3. Capping of radicals by molecular hydrogen and by hydrogen abstraction from the solvent inhibiting polymerization of intermediate radical species.

4. Thermal instability of coal derived materials leading to interconvertibility between products including retrograde reactions and the establishment of steady state conditions. The principle of reversibility is of importance in these processes.

5. The role of the catalyst at various stages of the process. Principally in coal dissolution reactions, in stabilization of reactive radical intermediates against polymerization, in promotion of bond cleavage reactions and in facilitating the transfer of hydrogen to coal-derived materials.   It is most unlikely that any single catalyst will have significant activity in all of these steps.

On the basis of our present knowledge of the role of iron- and tin-based catalysts and of the role of the sulphided nickel molybdate catalyst the mechanism shown in Figure 4 is proposed to summarize the essential steps in the hydroliquefaction of low rank coals.

At temperatures >360°C brown coal is rapidly dissolved in the vehicle without significant chemical reaction forming coal' (reaction 1) which corresponds to the pyridine soluble material observed by Neavel (11).   At increasing reaction time coal' is thermally converted into reactive radical intermediates (reaction 2) which in the absence of a suitable hydrogen donor, either molecular hydrogen or the vehicle, rapidly repolymerizes (reactions 3,4,5).   In the presence of a good hydrogen donor, hydrogen abstraction reactions (reactions 6,7) compete with polymerization reactions leading to the formation of oils and asphaltenes.   The important concept of reversibility is included in this scheme by the two-way arrows between the radical pool and the respective products.   It is possible therefore for polymerized material to re-react (reactions 2,8,9) and form lower molecular weight oils and asphaltenes.   Thus the system reaches a dynamic equilibrium after a period of time which is primarily dependent on the reaction temperature, the concentration of available hydrogen, and the metal-based catalyst.

At temperatures greater than 400°C, the asphaltenes rapidly degrade (reaction 10) and can either repolymerize or, in the presence of hydrogen, form oil.   At higher temperatures (>425°C) thermal cracking of oils, asphaltenes and repolymerized products leads to increasing yields of hydrocarbon gases (reaction 12).

In the presence of a high pressure of hydrogen, tin metal facilitates the stabilization of radicals formed by initial coal depolymerization leading to the formation of preasphaltenes and asphaltenes (reactions 5 and 6).   The stronger pressure dependence of the tin catalyzed reactions compared with other systems investigated here may be linked

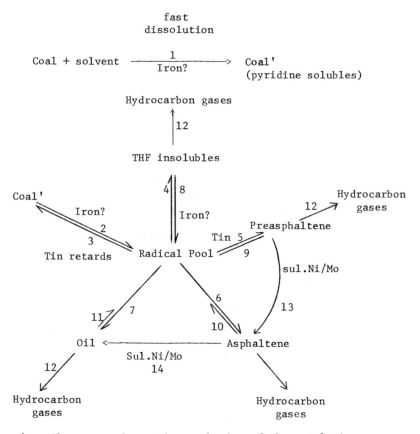

Figure 4   The proposed reaction mechanism of the catalytic
              hydroliquefaction of Victorian brown coal.

Note   The length of the arrows in reactions 8,9,10,11
              indicate the relative reactivities of the respective
              fractions.

to the fact that the amount of hydrogen dissolved in liquid tin is
also pressure dependent.

In the absence of tin less asphaltene is produced which manifests
itself as increased yields of THF insolubles and preasphaltenes. At
this time the activity of iron is not completely certain as no
effects have been observed on the reactions of the coal derived
products.   However the proven ability of iron to both catalyze the
reactions of some ethers while suppressing propagation reactions of
phenoxy radicals suggests that its major activity is probably
restricted to the first minutes of reaction (reactions 1,2) and to
the slow catalytic degradation of polymerized material.   The iron-
tin synergism (1) can be interpreted as a co-operative action

between the catalytic and radical stabilization activities of iron
with the hydrogen utilization and radical stabilization activities
of tin which allows the radical intermediates to be stabilized as
oil and asphaltene.

The conversion of asphaltene to oil is not catalyzed by either iron
or tin but it is facilitated by increased reaction temperature and
the availability of a good H-donor solvent. The resulting oil
formed at high temperatures (>400°C) contains much less oxygen and
the conversion probably is the result of the loss of polar phenolic
groups as well as cleavage of carbon-carbon bonds joining aromatic
clusters together. Vernon (15) has shown that dibenzyl thermally
decomposes above 400°C.

Formation of hydrocarbon gases principally results from thermal
cracking although tin catalyzes the formation of methane at
temperatures >425°C. Weller (16) has also shown that tin(II)
chloride catalyzes the formation of methane in the reaction of
1-methylnaphthalene with hydrogen.

The action of sulphided Ni/Mo catalyst was to dramatically increase
the yield of oil from the reactions of preasphaltene and asphaltenes. It is believed that the mechanism of the Ni/Mo catalyst
involves a more conventional dissociative adsorption of both
hydrogen and reactant molecule to the catalyst surface followed by
hydrogenolysis and hydrogenation reactions. The mechanistic path-
way thus differs from both iron and tin and is shown in Figure 4 by
reactions 13 and 14.

Acknowledgments

Financial support for this work was provided by the Victorian Brown
Coal Council and the Australian Government Department of National
Development and Energy under the National Energy, Research,
Development and Demonstration Program. The views expressed herein
are entirely those of the authors.

Literature Cited

1. Marshall, M., Jackson, W.R., Larkins, F.P., Hatswell, M.R.,
    and Rash, D., Fuel 61, 121 (1982).

2. Cassidy, P.J., Hertan, P.A., Jackson, W.R., Larkins, F.P., and
    Rash, D., Fuel 61, 939 (1982).

3. Hatswell, M.R., Jackson, W.R., Larkins, F.P., Marshall, M.,
    Rash, D., and Rogers, D.E., Fuel 62, 336 (1983).

4. Cochran, S.J., Hatswell, M., Jackson, W.R., and Larkins, F.P.,
    Fuel 61, 831 (1982).

5. Cassidy, P.J., Jackson, W.R., and Larkins, F.P., Fuel 62, 1404
    (1983).

6.   Hertan, P.A., Jackson, W.R., and Larkins, F.P., submitted to
     Fuel (1984).

7.   Agnew, J.B., Jackson, W.R., Larkins, F.P., Rash, D., Rogers,
     D.E., Thewlis, P., and White, R., Fuel 63,

8.   McPhail, I., and Murray, J.B., SECV Miscellaneous Report
     No. MR-155 31 March 1969.

9.   Weller, S., Clark, E.L., and Pelipetz, M.G., Ind. Eng. Chem.
     42, 334 (1950).

10. Schlosberg, R.H., Ashe, T.R., Pancirov, R.J., and Donaldson, M.,
     Fuel 60, 155 (1981).

11. Neavel, R.C., Coal Science 1, 1 (1982).

12. Whitehurst, D.D., and Mitchell, T.O., Am. Chem. Soc. (Div. Fuel
     Chem.) Preprints 26(2), 173 (1981).

13. Curran, G.P., Struck, R.T., and Gorin, F., Ind. Eng. Chem. Proc.
     Des. Dev. 6, 166 (1967).

14. Grandy, D.W., and Petrakis, L., Fuel 58, 239 (1979).

15. Vernon, L.W., Fuel 59, 102 (1980).

16. Chien, P-L., Sellers, G.M., and Weller, S.W., Fuel Proc. Tech.
     7, 1 (1983).

RECEIVED April 1, 1984

Structure and Liquefaction Reactions of Texas Lignite

C. V. PHILIP, R. G. ANTHONY, and ZHI-DONG CUI

Kinetic, Catalysis and Reaction Engineering Laboratory, Chemical Engineering Department, Texas A&M University, College Station, TX 77843

Texas lignite samples freshly collected from the mines were liquefied using mini-reactors. The lignite conversion products were characterized in order to study the lignite liquefaction reactions as well as the structural details of Texas lignite. The gas samples from the reactors were analyzed by the use of totally automated multi-column gas chromatographs with switching valves. The tetrahydrofuran (THF) solubles were separated by gel permeation chromatography (GPC). The GPC fractions were collected and analyzed using a gas chromatographic system equipped with a wide bore fused silica capillary column and with an ionization detector. The analytical data are used to illustrate the structure and the direct liquefaction reactions of Texas lignite.

The conversion of low-rank coals such as lignites into gaseous and liquid fuels and chemical feed stocks received considerable attention in the last decade (1-3). Data for liquefaction and gasification of various lignites have been reported (1,3-16). Although extensive analytical data are available on lignite conversion products, the use of those data to monitor chemical reactions involved in any liquefaction or gasification process is limited due to the enormous amount of time and effort required for the complete characterization (17-21). In the case of most of the reports on liquefaction experiments, the data are usually limited to total conversion amount of certain fractions such as oils, asphaltenes, preasphaltenes, tetrahydrofuran solubles and certain physical characteristics (17). This study involves one lignite sample from a Texas mine and its conversion products from mini-reactor experiments using varying liquefaction conditions. The gaseous products are analyzed by

0097-6156/84/0264-0287$06.00/0
© 1984 American Chemical Society

gas chromatography (GC) (14) and the THF solubles are
characterized by gel permeation chromatography (GPC) (8-15) and
high resolution gas chromatography (HRGC).

Experimental

Lignite samples were collected fresh from the mine near Carlos,
Texas (Jackson group) and stored under distilled water prior to
use. The lignite was ground to less than 20 mesh size and
stored in a closed jar which was kept in the refrigerator. The
lignite liquefaction experiments were conducted in four 6.3 ml
mini-reactors which were heated in a fluidized sand bath. The
reactors were fabricated using "Autoclave" fittings. Each reactor was the same, but only one reactor was fitted with a pressure transducer. The absence of pressure gauges on the other
three reactors allowed the minimization of the dead volume
outside the sand bath. In most cases 1.5 gm lignite, 1 ml solvent (decalin or tetralin) and a gas (nitrogen, hydrogen or 1:1
CO/H_2 mixture at 1000 psi) were charged to the reactor. The
reactions were quenched by dipping the reactors in cold water.
The gases from the reactors were analyzed using an automated
multi-column gas chromatograph equipped with sample injection
and column switching valves (4). The contents of the reactors
were extracted with THF by using an ultrasonic bath. The
THF extract (about 15 ml) was filtered through a 1 micron micropore filter (Millipore) and separated into fractions by using a
gel permeation chromatograph (GPC) (15).

The GPC fractions were obtained with a Waters Associate
Model ALC/GPC 202 liquid chromatograph equipped with a refracto-
meter (Model R401). A Valco valve injector was used to load
about 100 μl samples into the column. A 5 micron size 100 Å PL
gel column (7.5 mm ID, 600 mm long) was used. Reagent grade
THF, which was refluxed and distilled with sodium wire in a
nitrogen atmosphere, was used as the GPC carrier solvent. A
flow rate of 1 ml per minute was used. THF was stored under dry
nitrogen, and all separations were conducted in a nitrogen
atmosphere to prevent the formation of peroxides.

Straight chain alkanes from Applied Science, aromatics from
Fisher Scientific Company and polystyrene standards from Waters
Associate were used without purification for the linear molecular size calibration of the GPC. Since the solubility of the
larger alkanes in THF is very low, approximately 0.2-1 mg of
each standard was dissolved in 100 μl of THF for the molecular
size calibrations. The fractions of the lignite derived liquid
separated by GPC were analyzed by a gas chromatographic system
(VISTA 44, Varian Associates) equipped with a 15 m long and
32 mm ID bonded phase (DB-5) fused silica capillary column and
a flame ionization detector (FID).

Results and Discussion

The lignite samples from Carlos, Texas have heating values of
4000 to 6000 Btu/lb on an as-received basis or about 12000
Btu/lb on dry mineral-matter-free (dmmf) basis. The samples may

contain as much as 40% moisture which escapes from lignite even while stored in the refrigerator (water condenses on the walls of the container). The oxygen content varied from 20 to 30% and has about 1.5 to 2% sulfur on a dmmf basis. As much as 50% of the oxygen may exist as carboxylic groups which may produce carbon dioxide at the liquefaction conditions (5). The lignite sample used in this steady had 35.7% moisture and the proximate analysis gave 39.5% volatiles, 23.7% fixed carbon and 36.7% ash on a dry basis. Elemental analysis gave 64.8% carbon, 6.1% hydrogen, 1.3% nitrogen, 1.7% sulfur and 26% oxygen (by difference) on a dmmf basis.

The substantial amount of hydrogen sulfide that was usually liberated decreased rapidly as a function of sample aging and storage conditions. The lignite seams are soaked with water under hydrostatic pressure in a virtually anaerobic condition. Immediately after mining, the lignite samples continuously lose moisture and are oxidized by air. There is no ideal way the lignite samples can be stored to preserve its in seam characteristics.

The lignite liquefaction conditions are listed in Table 1. The composition of gaseous products are also listed in Table 1. The data in Table 1 illustrate that the composition of gaseous products are unaffected by feed gas or the solvent system (whether hydrogen donating or not). Higher temperatures favor the production of hydrocarbons. Apparently the fragile species produced by the pyrolytic cleavages of bonds in the coal structure are converted to stabler species which form the components of lignite-derived gases without consuming hydrogen from the gaseous phase or from the hydrogen donor solvents. The species which require hydrogen for stabilization are abstracting hydrogen from carbon atoms in the vicinity. It is quite possible that the bond breaking and hydrogen abstraction are simultaneous and hydrogen from the donor solvent or the gaseous phase may not reach the reaction site in time to hydrogenate the fragile species.

To illustrate the gaseous hydrocarbon production from lignite, experiments were conducted using butylated hydroxytoluene (BHT), a widely used antioxident and food preservative, as a model compound (see Table 1). BHT decomposed to products including isobutylene and isobutane in an approximate 1:1 molar ratio in the absence of hydrogen from hydrogen donor solvent or gaseous phase. Addition of tetralin favored the production of isobutane over isobutylene. It appears that lignite can give some hydrogen to the isobutane formation but tetralin can readily reach the reaction vicinity in time for the hydrogen transfer reaction.

The elution pattern of the GPC using 5 micron 100 Å PL gel column is illustrated in Figure 1 where the GPC separation of a standard mixture containing straight chain alkanes and aromatics is shown. The polystyrene standard (mol. wt. 2350 and chain length 57 Å) gave a broad peak at 11 ml retention volume. The peak position is marked in the figure rather than using polystyrene standard in the mixture in order to save the $nC_{44}H_{80}$ peak from the enveloping effect of the broad polystyrene peak.

TABLE 1

Run No.	Lignite Liquefaction Conditions				Composition of Product Gas		
	Solvent 1 ml	Feed Gas (1000 psi)	Temp °F	Time min.	H_2	CO_2	CO
1	tetralin[**]	N_2	700	30	13.0	11.2	26.1
2	tetralin[**]	N_2	700	120	2.1	61.5	3.6
3	tetralin	N_2	700	120	0.9	80.0	0.9
4	decalin[**]	N_2	700	120	0.3	61.9	2.1
11	tetralin	$CO+H_2(1:1)$	700	120	3.2	60.4	2.5
12	tetralin	N_2	700	120	3.7	70.5	2.8
13	decalin	N_2	700	120	0.5	75.9	3.8
14	tetralin	N_2	700	120	2.3	78.2	2.9
21	tetralin	N_2	700	120	1.1	80.7	1.5
22	tetralin	none	700	120	5.2	51.2	3.4
23	none	none	700	120	-	76.5	3.6
24	tetralin	N_2	700	120	0.8	83.7	1.6
31	tetralin	H_2	700	120	0.7	71.3	2.7
32	tetralin	$CO+H_2(1:1)$	700	120	0.7	70.1	1.9
33	decalin	H_2	700	120	0.7	75.5	2.1
34	decalin	$CO+H_2$	700	120	0.8	63.0	2.4
41	tetralin	N_2	700	15	-	80.8	2.1
42	tetralin	N_2	700	120	1.0	74.3	1.5
43	tetralin	N_2	700	360	5.6	41.2	1.3
51	tetralin	N_2	850	30	8.8	44.7	4.7
52	tetralin	H_2	850	30	8.1	46.6	5.3
53	tetralin	$CO+H_2(1:1)$	850	30	9.9	47.5	5.4
61	tetralin	N_2	800	30	1.7	65.8	6.0
62	tetralin	H_2	800	30	1.6	65.8	6.0
63	tetralin	$CO+H_2(1:1)$	800	30	1.6	66.1	5.9
71	tetralin	N_2	750	30	0.7	78.4	3.7
72	tetralin	H_2	750	30	0.7	76.9	2.4
73	tetralin	$CO+H_2$	750	30	0.7	75.2	2.9

[*] Product distribution does not include the feed gas, i.e., feed gas free basis

[**] + 250 mg BHT

Composition of Product Gas,[*] mole %

H_2S	CH_4	C_2	$C_2^=$	C_3	$C_3^=$	$n-C_4$	$i-C_4$	$i-C_4^=$
-	8.4	1.3	1.9	0.8	0.15	0.07	12.6	24.1
-	9.2	2.8	2.8	1.8	0.28	0.19	11.9	5.9
-	11.8	3.7	7.3	1.2	0.13	0.17	0.1	0.03
-	8.6	2.0	1.5	1.0	0.10	0.07	8.7	13.2
22.4	6.8	2.6	0.2	1.0	0.12	0.14	0.1	0.01
10.5	7.9	2.6	0.3	1.0	0.14	0.15	0.1	0.1
7.8	8.4	2.2	0.6	0.6	0.10	0.06	0.04	0.02
1.1	9.7	3.4	1.2	1.1	0.10	0.13	0.1	0
-	9.5	2.6	0.3	1.5	0.15	0.16	1.8	0.8
-	28.1	3.6	0.04	3.2	0.01	0.06	3.9	0.9
-	11.4	3.9	0.2	2.0	0.31	0.41	0.2	0.1
-	8.3	3.1	0.3	1.2	0.18	0.17	0.1	0.03
13.1	7.3	2.6	0.7	0.9	0.12	0.18	0.2	0.2
12.7	5.3	1.8	0.4	0.6	0.10	0.09	0.1	0.1
11.3	7.3	1.9	0.1	0.7	0.12	0.07	0.1	0.03
-	19.1	4.2	1.1	2.2	1.64	0.90	1.3	1.1
11.5	3.9	0.8	0.3	0.2	0.09	0.03	0.03	0.03
10.2	7.5	3.2	0.8	1.1	0.13	0.15	0.1	0.02
37.1	9.2	3.7	0.1	1.4	0.11	-	0.1	0.01
8.8	21.0	5.5	0.4	2.2	0.26	0.44	0.2	-
7.6	23.2	6.6	0.4	1.7	0.23	0.17	0.1	0.02
9.9	19.0	5.5	0.1	1.8	0.24	0.29	0.2	-
9.4	11.1	4.3	1.2	0.1	0.23	0.21	0.1	0.03
8.8	11.8	3.9	0.6	1.0	0.16	0.11	0.1	0.02
10.4	11.5	3.2	0.1	0.9	0.14	0.09	0.1	0.02
4.9	8.3	2.4	0.4	0.7	0.14	0.11	0.1	0.03
6.6	9.2	2.9	0.1	0.8	0.14	0.10	0.1	0.02
8.7	8.0	2.8	0.5	0.8	0.16	0.10	0.1	0.02

The retention volume of several aliphatic, phenolic, heterocyclic, amine and aromatic compounds in THF and toluene have
been reported elsewhere (9). It is clear that aromatic compounds, as expected from their valence bond structures, have
smaller linear molecular sizes compared to n-alkanes of similar
molecular weight (15). It is expected that most of the condensed ring aromatics such as naphthlene, anthracene and even
big ones like coronene (seven fused rings with molecular weight
of 300.4) are smaller than n-hexane and hence have retention
volumes larger than that of n-hexane. The polystyrene standard
with a 57 Å appears to be larger than what is expected from a
comparable alkane. The large number of phenyl groups on the main
polyethylene chain (57 Å) causes the molecule to behave like a
rigid cylindrical structure with large steric hindrance for
penetrating the pores of the gel. The two terminal phenyl
groups also contribute to an increased chain length. The
polystyrene peak (57 Å) is very close to the total exclusion
limit of the 100 Å PL gel column.

The THF solubles of lignite liquefaction products were
separated by GPC on the basis of effective linear molecule sizes
in solution (15). Although GPC can be used for molecular weight
or molecular volume separation of homologous series such as
polymers, such separations of complex mixtures like coal liquid
is not feasible. The only molecular parameter which has the
least variation from a calculated value is effective linear
molecular size in solution. Lignite-derived liquids are
separated on the basis of effective molecular length which is
expressed in carbon numbers of straight chain alkanes. The GPC
fraction with species larger than $nC_{44}H_{80}$ is in fraction 1,
which is composed of nonvolatile species. Fraction 2 has
molecular sizes in the range of nC_{14} to nC_{44} and composed of
volatile species, mostly alkanes and nonvolatile species
generally known as asphaltenes. Fraction 3 (nC_7 - nC_{13}) is
composed of alkylated phenols such as cresols, alkyl indanols
and alkyl naphthols as well as some small amount of nonvolatiles
namely low molecular weight asphaltenes. Fraction 4 is composed
of species with molecular size less than that of $n-C_7H_{16}$. This
fraction may not contain any straight chain alkanes as they are
very volatile. It is composed of aromatic species such as
alkylated benzenes, alkylated indans and naphthalene and even
large species such as pyrenes and coronenes. The solvent system
used for liquefaction (tetralin and decalin) separate from the
bulk of the lignite- derived products as the last peak (peak at
20.5 ml). Although the column is overloaded with respect to the
solvent system, the efficiency of separation of the lignite-
derived products is unaffected. The volatile species of all
fractions can be identified by GC-MS(15) and nonvolatiles by IR
and NMR spectroscopy (12,13,17).

The effect of hydrogen donor solvent and feed gases such as
hydrogen and CO/H_2 mixture on lignite dissolution at 700°F is
shown in Figure 2. Decalin dissolves less coal compared to
tetralin. CO/H_2 as feed gas gave less liquid products which
contained more larger molecular size species. When decalin with
nitrogen (not shown in Figure 2) was used for liquefaction of

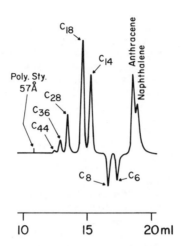

Figure 1. GPC of calibration mixture.

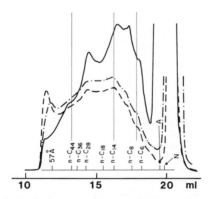

Figure 2. GPC of THF solubles of lignite from reactions at 700 °F in decalin and H_2 (Run 33) (——), tetralin and H_2 (Run 31) (---), and tetralin and CO/H_2 (Run 32) (-·-).

lignite the species appearing at total size exclusion limit of 100 Å gel column were totally absent. Figure 3 shows GCs of fraction 2 of two lignite-derived liquids to illustrate the role of the hydrogen donor solvent on alkane production. If the alkanes are formed by breaking of alkyl chains from other groups, they are expected to contain more olefinic species in the absence of hydrogen donor solvents such as tetralin. Since the relative concentration of species in Figures 3 a and b are very similar with the exception of pristane, the role hydrogen or hydrogen donor solvent plays in the production of alkanes is limited. Simple pyrolysis can liberate them from the coal matrix. The alkane may be existing as free while trapped inside the pores or their precursors such as carboxylic acids or species which decompose during heating to produce alkanes easily. The release of pristane, a branched alkane ($C_{19}H_{40}$), is enhanced by hydrogen donor solvent.

Figure 4 illustrates the effect of reaction time on liquefaction. Figures 5 and 6 show the liquefaction of lignite at $750^{\circ}F$ and $800^{\circ}F$ using different feed gases. Hydrogen tends to give more phenols and aromatics compared to nitrogen. CO/H_2 tends to produce higher molecular size species at the expense of phenols and aromatics. A qualitative estimation of the solvent systems can be obtained by estimating the width of the solvent peak, and the area of the rest of the GPC may give the dissolved product from lignite. The use of CO/H_2 did not increase the total liquefaction yield while it has decreased the total yield of liquid products. The experiments using CO/H_2 gave liquid products which were difficult to filter through micropore filters compared to the products obtained from experiments using either nitrogen or hydrogen as the feed gas. It could be assumed that CO/H_2 will give a product which contained more large size asphaltenes.

Structure of Lignite

The structure of coal can best be understood by examining the plant tissues and the possible pathways for degradation of plant tissues during the coalification process rather than just looking at the coal and coal-conversion products. It could be generalized that plants contain about 50% cellulose, 25% lignin and 25% of a number of substances such as proteins, waxes, terpenes and flavonoids (22). Cellulose may undergo a dehydration process and lignin may undergo a deoxygenation process during coalification. Dehydrated cellulose (as well as other carbohydrates) and partially deoxygenated lignin may form the principal skeletal structure of coal. Other constituents of plants may also undergo chemical rearrangement as they form part of the coal structure. Chlorophyll and plant proteins are the major nitrogen compounds in plants, and they could be transformed into precursors which cause the formation of heterocyclic nitrogen compounds and alkylated amines in coal conversion products. Dehyrated cellulose may give rise to carbon skeletons which may not be converted to hydrocarbons during coal liquefaction. Some of the plant constituents such as cellulose and

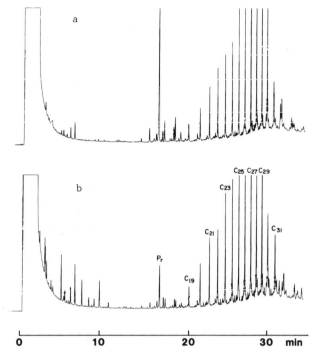

Figure 3. GC of alknae fractions from THF solubles of lignite. Key: a, in tetralin and H_2 (Run 21); and b, in decalin and N_2 (Run 13).

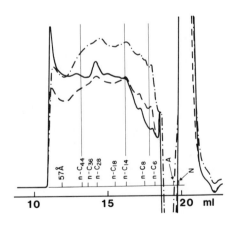

Figure 4. Effect of reaction time on THF solubles of lignite in tetralin and H_2 at 700 F. Key to time (min): ———, 15 (Run 41); ———, 120 (Run 42); —·—, 360 (Run 43).

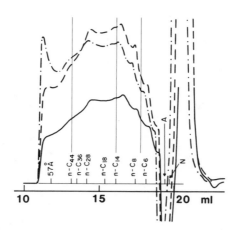

Figure 5. Effect of feed gas on lignite liquefaction in tetralin at 750 °F. Key to feed gas: ——, N_2 (Run 71); ---, H_2 (Run 72); and –·–, CO/H_2 (Run 73).

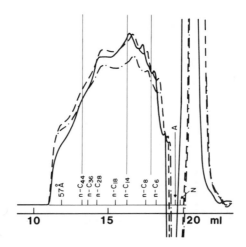

Figure 6. Effect of feed gas on lignite liquefaction in tetralin at 800 °F. Key to feed gas: ——, N_2 (Run 61); ---, H_2 (Run 62); and –·–, CO/H_2 (Run 63).

proteins are susceptible to microbial attack, and carbon and nitrogen may be lost as end products of biological degradation in the form of carbon dioxide and ammonia, while substances like lignin may be unaffected. Hence the coal may have a higher fraction of lignin derived product than previously expected from the plant precursors.

During liquefaction, gases such as CO_2, H_2S, CO, CH_4, C_2H_6, C_3H_8 and other small hydrocarbons are produced. The compositions of a few samples of lignite-derived gases have been reported elsewhere (4). Ten to 15% of lignite (dmmf) is converted to carbon dioxide (i.e., one mole of carbon dioxide is produced per two moles of lignite based on an empirical formula calculated from the elemental analysis). The methane produced amounts to a molecule per five empirical formulas and an ethane molecule per every twelve formulas. The production of isobutane and n-butane is small compared to that of methane, ethane and propane. This may be due to the fact that methyl, ethyl and propyl groups are present more abundantly than butyl or isobutyl groups. It is interesting to note that the CO amounts to only one molecule per 30 formulas. Only a small quantity of the low boiling liquid is produced, and its analysis by GC and GC-MS has been reported (13). The products of most significance are the liquefaction products known as synthetic crude. The GPC separation of Texas lignite-derived fluids and analysis of GPC fractions by GC-MS have been reported (15). Figure 2 illustrates how a lignite derived-liquid sample could be separated into five fractions with different chemical species. Fraction 1 is composed of nonvolatile species with linear molecular sizes larger than $nC_{44}H_{90}$ which is the largest alkane that could be detected by the GC system used for the analysis. Fraction 1 represents only a small fraction of the liquefaction products. Fraction 2 has volatile components--alkanes ranging from $C_{14}H_{30}$ to $C_{44}H_{90}$-- as well as nonvolatile compounds which could be called asphaltenes. The asphaltenes could be separated also by vacuum distillation or by dissolving asphaltenes selectively in liquid sulfur dioxide (6). Fraction 3 contains small amounts of nonvolatiles with linear molecular size less than $C_{14}H_{30}$ and a series of alkylated phenols, alkylated indanols and alkylated naphthols. A number of isomers are possible by adding CH_3 to C_5H_{11} to the parent nucleus of each species, and a large number of isomers are detected by GC-MS. Fraction 4 is composed of small amounts of smaller size phenols and the hydrocarbons representing the deoxygenated forms of phenols identified in fraction 3 such as alkylated benzenes, alkylated indans and alkylated naphthalenes. There were more of the alkylated phenols in the phenolic fraction while more of the aromatic fraction was composed of alkylated indans. Benzene and indan are the two predominant structures present in any coal liquefaction product. Fraction 5 is composed of tetralin and naphthalene. Although methylindan is an impurity in the technical grade tetralin, its concentration did not change appreciably.

The THF extract of lignite yields species in the higher
molecular size range but the GC does not show any volatiles such
as the alkanes (6).

The samples of synthetic crude produced from West Virginia
(10) subbituminous coal and North Dakota lignite are composed of
species similar to those found in Texas lignite-derived products
except for the polyaromatic hydrocarbons with three or more
aromatic rings, such as phenanthrenes, pyrenes, and benzo-
pyrenes. These polyaromatic hydrocarbons (PAH) could be produced in the pilot plants that were operated at higher temperatures (450°C) compared to the low temperatures (375°C) used for
Texas lignite liquefaction. The absence of PAH as well as
heterocyclic compounds such as benzothiophenes and quinolines in
coal has been reported (23). The Texas liquefaction products
contain only trace amounts of heterocyclic compounds. Liquid
sulfur dioxide extracted 82% of the SRC from West Virginia
subbituminous coal and 74% of the SRL from North Dakota lignite
(8). Liquid sulfur dioxide dissolves only the asphaltenes,
phenols and aromatics (8) while alkanes, minerals and colloidal
carbon are insoluble. The extract of SO_2-insoluble in THF is
8.5% and 9% in the case of SRC and SRL, respectively. The THF
extract contains alkanes as well as polymeric species composed
mostly of carbon skeletons. The residue of the SRC and SRL
after THF extraction has 16% ash content with 84% composed of
high carbon/low hydrogen species which could have been formed by
the dehydration of cellulose (11).

Liquefaction of Texas lignite produces volatile species
which have a high hydrogen to carbon ratio compared to the
starting material (7). All the hydrogen increase in the volatile product could not be accounted for by the donor-hydrogen
from tetralin. The analysis on asphaltenes from SRC (the SO_2-
solubles but nonvolatiles) has a hydrogen to carbon ratio of
close to 1. Hence, during liquefaction, components with low
hydrogen to carbon ratio are released from the coal matrix.
These species could have been formed by the dehyration of cellulose as well as other carbohydrates present in the plants which
survived microbial attacks as illustrated in Figure 7. The
phenols and the aromatics produced during liquefaction represent
the partially decayed lignin structures which appear to be
intact as far as the cyclic structures are concerned. The
alkanes found in liquefaction products could have been derived
from plant sources such as terpenes, oils and chlorophyll (pristane and phytane). The coal may have a structure made of partially decomposed lignin and dehydrated cellulose (skeletal
carbon with low hydrogen content). The residues from chlorophyll and proteins may account for the species containing
nitrogen, which may also exist as polymers.

During lignite liquefaction mostly the lignin type structures are cleaved. The ether links of lignin polymers are
surviving the long coalification process. The rupture of these
ether bonds aided by donor-hydrogen yields phenols and aromatics
in a ratio 1:1 in the case of Texas lignite. During the
cleavage, the formations of alkyl hydroxybenzenes (phenols) and
alkylindans are slightly favored over alkylindanols and alkyl-

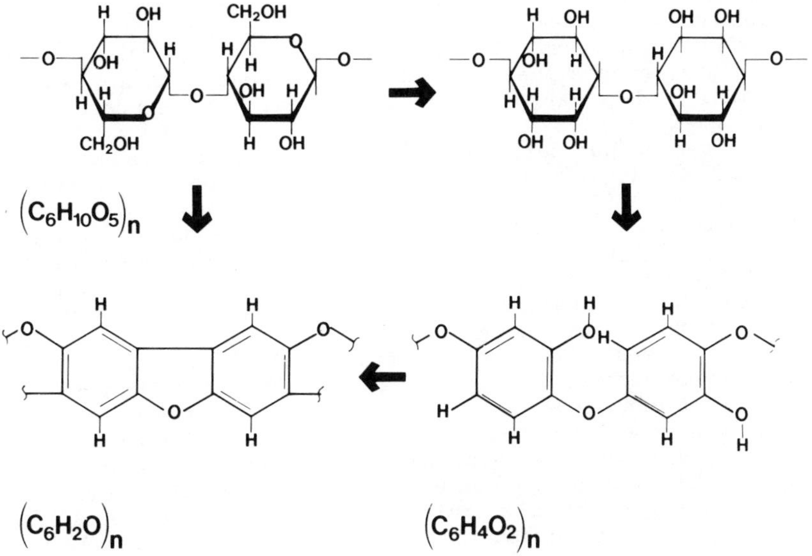

Figure 7.   Coalification of cellulose.

benzenes. The carbon skeletal structures with low hydrogen
content are aromatic in nature, and the smaller ones are soluble
in solvents like THF.

The structure of lignite may be visualized as being similar
to reinforced concrete. The dehydrated cellulose resembles the
steel, decomposed lignin is the cement, and the other species
such as alkanes play the role of sand. During coalification,
some inorganic substances will also migrate into the coal. A
general structural model for coal is presented in Figure 8. Low-
rank coals such as lignite have a high oxygen content of about
25% by weight on a dmmf basis, almost half of which exists as
carboxylic groups. Higher rank coals have lower oxygen content
which could have been achieved by decarboxylation during the
coalification process. The structure does not show all types of
structural groups present in a coal macrostructure. Only a
fragment of a large three-dimensional coal structure is
visualized.

Based on our data on lignite liquefaction products and
various coal liquids and their distillates, we proposed a structure for coal (7) as shown in Figure 8. The major structural
constituents of coal are derived from three sources namely
cellulose, lignin and other plant components dispersed in the
plant tissues. The first two are polymers. The coalification
process, which is mostly a deoxygenation process, might not
create a sufficiently large number of tertiary bonds needed for
binding various constituents of coal together to form a strong
three- dimensional structure. The coals, including lignite, may
have loose structures. The cellulose-derived structures may
have large pores in which species such as large alkanes could be
trapped. Although low temperature (less than 300°F) extractions
can extract some soluble components of coal, the alkanes are not
liberated (5) except as a small amount of carboxylic acids. The
low temperature oxidation studies do not detect large alkanes
(23). The higher temperatures (above 650°F) can either break
the pore structures or the alkanes can distill out of the pores.

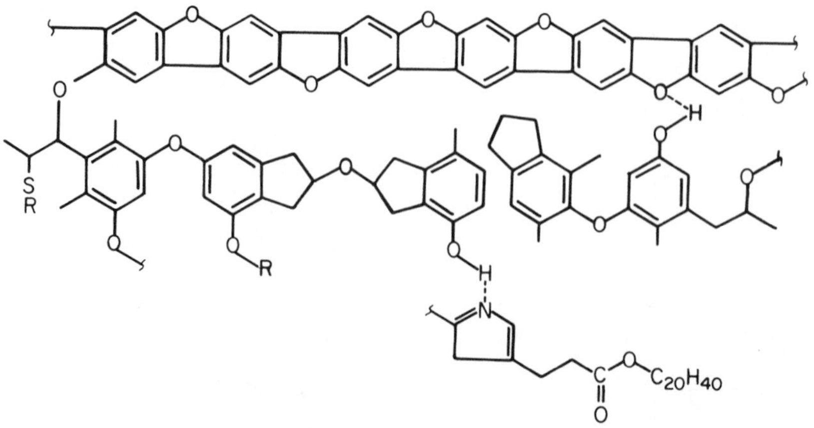

Figure 8. A structural model of coal.

Acknowledgments

The financial support of The Texas Engineering Experiment Station, The Texas A&M University Center for Energy and Mineral Resources, is very much appreciated. The lignite was supplied by Texas Municipal Power Agency (TMPA), Carlos, Texas.

Literature Cited

1. Sondreal, E. A.; Wilson, W. G.; Sternberg, V. I. Fuel 1982, 61, 925.
2. Proc. Low-Rank Coal Technology Development Workshop, San Antonio, Texas, 1981 (June 17-18).
3. Proc. Basic Coal Science Workshop, Houston, Texas, 1981 (Dec. 8-9).
4. Philip, C. V.; Bullin, J. A.; Anthony, R. G. J. Chromatogr. Sci. 1979, 17, 523.
5. Philip, C. V.; Anthony, R. G. Fuel Processing Technology 1980.
6. Philip, C. V.; Anthony, R. G. Proc. Coal Technology 1978, 2 p. 710.
7. Philip, C. V.; Anthony, R. G. Preprints ACS Org. Coat & Polym. Div. 1980 (August 20).
8. Zingaro, R. A.; Philip, C. V.; Anthony, R. G.; Vindiola, A., Fuel Processing Technology 1981, 4, 169.
9. Philip, C. V.; Anthony, R. G. Am. Chem. Soc. Div. Fuel Chem. Preprints 1979, 24, (3), 204.
10. Philip, C. V.; Zingaro, R. A.; Anthony, R. G. Am. Chem. Soc. Fuel Chem. Preprints 1980, 25, (1), 47.
11. Philip, C. V.; Zingaro, R. A.; Anthony, R. G. in "Upgrading of Coal Liquids"; Sullivan, R. F., Ed.; ACS SYMPOSIUM SERIES No. 156, American Chemical Society: Washington, D.C., 1981; p. 239.
12. Philip, C. V.; Anthony, R. G. Am. Chem. Soc. Div. Fuel Chem. Preprints 1977, 22, (5), 31.
13. Philip, C. V.; Anthony, R. G. in "Organic Chemistry of Coal"; American Chemical Society: Washington, D.C., 1978, p. 258.
14. Philip, C. V.; Anthony, R. G. Fuel 1982, 61, 351.
15. Philip, C. V.; Anthony, R. G. Fuel 1982, 61, 357.
16. Farnum, S. A.; Olson, E. S.; Farnum, B. W.; Wilson, W. G.; Am. Chem. Soc. Div. Fuel Chem. Preprints 1980, 25, 245.
17. Aczel, T.; Williams, R. B.; Pancirov, R. J.; Karchmer, J. H. "Chemical Properties of Synthoil Products and Feeds," U.S. Energy Research and Development Administration, 1976, FE8007.
18. Baltisberger, R. J.; Klabunde, K. J.; Sternberg, V. I; Wolsey, N. F.; Saito, K.; Sukalski, W. Am. Chem. Soc. Div. Fuel Chem. Preprints 1977, 22, (5), 84.
19. White, C. M.; Sulz, J. L.; Sharkey, A. C., Jr. Nature 1970, 268, 620.

20. Pugmire, R. J.; Grant, D. M.; Zilm, K. W.; Anderson, L. L.;
    Oblad, A. G.; Wood, R. E. Fuel 1978, 57, (4).
21. Seshadri, K. S.; Albaugh, E. W.; Bacha, J. D. Fuel 1982,
    61, 336.
22. Lehninger, A. L. "Biochemistry," Worth Publishers: New
    York, 1970; p. 231.
23. Duty, R. C.; Hayatsu, R.; Seott, R. G.; Moore, L. P.;
    Winams, R. E.; Studier, M. H. Fuel 1980, 59, 97.

RECEIVED March 12, 1984

INDEXES

Author Index

Allardice, D. J., 3-14
Anthony, R. G., 287-302
Bale, H. D., 79-94
Benson, S. A., 39-52,175-94
Carlson, M. L., 79-94
Cassidy, P. J., 275-86
Chaffee, A. L., 109-32
Cui, Z.-D., 287-302
Diehl, E. K., 195-210
Edgar, T. F., 53-76
Farnum, B. W., 145-58
Farnum, S. A., 145-58
Hengel, T. D., 267-74
Hertan, P. A., 275-86
Huffman, G. P., 159-74
Huggins, F. E., 159-74
Jackson, W. R., 275-86
Jenkins, R. G., 213-26
Johns, R. B., 109-32
Kaiser, W. R., 53-76
Kalliat, M., 79-94
Karner, F. R., 39-52,175-94
Kiss, L. T., 3-14
Kobylinski, T. P., 227-42
Kube, W. R., 39-52
Kwak, C. Y., 79-94

Larkins, F. P., 275-86
Mallya, N., 133-44
Marshall, M., 275-86
Miller, D. J., 145-58
Morgan, B. A., 255-66
Morgan, M. E., 213-26
Perry, G.J., 3-14
Philip, C. V., 287-302
Rash, D., 275-86
Roaldson, R. G., 175-94
Scaroni, A. W., 255-66
Schmidt, P. W., 79-94
Schobert, H. H., 39-52,175-94,195-210
Setek, M., 95-108
Snook, I. K., 95-108
Streeter, R. C., 195-210
Timpe, R. C., 145-58
Verheyen, T. V., 109-32
Wagenfeld, H. K., 95-108
Walker, Jr., P. L., 267-74
Watkins, I. D., 275-86
Wen, C. S., 227-42
Wolfrum, E. A., 15-38
Young, B. C., 243-54
Zingaro, R. A., 133-44

Subject Index

A

AA--See Atomic absorption
Absaloka, computer-controlled SEM
    results, 164t
Absorption mode IR spectroscopy, 114
Acetic acid
  derivation, 117
  formation, 138
Acid(s), in lithotypes, 122t
Acid fractions, 123
  separation, 121f,122f
Acidity
  determination, 134-35
  Wilcox lignite, 140t

Acyclic hydrocarbons,
    analysis, 147-53,149t
AFBC--See Atmospheric fluidized-bed
    combustion
Alcohols, composition in
    lithotypes, 122t
Aliphatic chains, Wilcox lignite, 141
Alkane(s), detection, 153
n-Alkane(s)
  vs. depth, carbon preference
      indices, 127f
  distributions, 124
    extract, 151-52f
  extracted/separated, 150t
  liquefaction products, 149t

305

Alkane production, hydrogen donor
  solvent, 294
Alkylbenzenes, detection, 153
Aluminum
  and precipitator ash, 14
  in Kinneman Creek bed, 186f
Aluminum oxide, in coal combustion, 14
Anhydrite, in Montana lignite, 217t
Anthracite, kinetic data, 249t
Antimony, position in lignite
  seam, 188
Aromaticity, 41
Ash
  Beulah lignite, 89t
  bituminous coals, 41t
  northern Great Plains lignites, 40t
  Rhenish brown coals, 21
  sulfur retention, 68
  Texas lignite, 60-61t,64-65t
  tracer technique, 260
  Victorian brown coal, 6t
Asphaltenes
  conversion to oil, 285
  degradation, 283
  separation, 297
Atmospheric fluidized-bed combustion,
  Texas lignite, 70-71
Atomic absorption, Victorian brown
  coal, 4-5
Australian subbituminous coals
  See also Victorian brown coals
  chars
    properties, 246t
    uses, 244
  combustion reactivity, 243-54
Azeotrope-soluble extracts, 147t,148f
Azeotropic distillation, 4

B

Barium, in Montana lignite, 216t
Bassanite, formation, 62
Bed-moist conditions, 107
Bed-moisture content, 4
  Victorian brown coal, 6t
Beulah lignite
  acyclic hydrocarbons, 149t
  n-alkane extract distribution, 151f
  analyses, 89t,146t
  chloroform extracts, 145-58
  hexane-soluble extracts, 150t
  scattered intensity, 86f
  sesquiterpene distribution, 153
  X-ray specific surfaces, 88
BHT--See Butylated hydroxytoluene
Big Brown, analyses, 145-58,146t
Bishomohopanoic acid, ratio of
  diastereomers, 129f

Bituminous coals
  analyses, 40t
  ash composition, 41t
  computer-controlled SEM, 161
  mineralogy, 167t
  vs. northern Great Plains
    lignites, 40-41
  organic structural relationships, 41
Boiler fouling
  sodium effects, 13
  with Texas lignite, 69
Boiler-wall slag deposits, Mossbauer
  spectrum, 172f
Briquetting, Rhenish brown
  coals, 28-30
Brown coal
  dry
    microstructure
      determination, 95-108
    SAXS, sample preparation, 98
  gas adsorption data, 104t
  lithotypes
    heating value, 24f
    pertrifluoroacetic acid
      oxidized, 118t
  polycyclic aromatic
    hydrocarbons, 128
  Rhenish, 15-38
    ash content, 21
    briquetting, 28-30
    characterization, 16,19f,20f,21
    chemo-physical
      characterization, 24f
    coking, 28-30
    composition, 16-21
    direct hydrogenation, 33
    gasification, 30-31
    heating value, 23,24f,25f
    huminite content, 21,22f
    humotelinite vs. lipto-
      humodetrite, 27f
    humotelite, 28
    hydrogasification, conversion, 32f
    hydrogen content, 23,28
    hydrogenation
      indirect, 31,32f,34f
      process engineering, 35
    hydrogenation coals,
      properties, 34f
    liptinite content, 21,22f
    liquefaction, 31,33
    oxygen functional groups,
      distribution, 18f
      quality vs. refining
        behavior, 28-30
    structure, 16-21,17f
    volatiles, 23,26f
  SAXS, 104t

Brown coal--Continued

Victorian
n-alkane distribution, 124
aromaticity vs. depth, 113f
ash content vs. minerals, 6t
atomic absorption, 4-5
bed-moisture content, 6t
calcium, 11f
carbon variation, 10f
carbonate formation, 13
chemical characteristics, 3-14
coal field variations, 8-12
color index, 7
extractable polycyclic aromatic
    hydrocarbons, 130f
n-fatty acids, 128
hydrogen index, 116f
hydrogen variation, 10f
hydroliquefaction, 275-86,279t
    mechanism, 284f
    pressure effects, 278f
    temperature effects, 277,278f
inorganic content, 4-5
IR absorption coefficients vs.
    depth, 115f
lithotypes, 8,10f
magnesium, 11f
mineral content, 4-5
moisture content, 4
Morwell
    Ca content, 8
    coal-ash chemistry, 7t
    depth variations, 9f
nonpyritic iron, 11f
oxygen functional groups, 7
oxygen index, 116f
preasphaltene hydrogenation, 282t
product distribution, 279t
properties, 4-8
    field variations, 12t
reaction with ether compounds, 280
Rock-Eval analysis, 116f
seams, variations, 7-8
sodium, 11f
water in clays, 5
X-ray diffraction, 13
X-ray fluorescence, 4-5
wet
    microstructure
        determination, 95-108
    SAXS, sample preparation, 98
Butylated hydroxytoluene, 289

C

Calcite, in Montana lignite, 217t
Calcite-rich bituminous coal
    phase-shift subtracted Fourier
        transform, 169f,170

Calcite-rich bituminous coal--Continued
    X-ray absorption near-edge
        spectra, 168f
Calcium, 257
    catalyst, char gasification, 268
    coexchanged chars, 270t
    distribution, Kinneman Creek
        Bed, 189f
    effects
        on combustion, 264,265f
        on pyrolysis, 262
        on Texas lignite, weight
            loss, 263f
    lignite char gasification, 267-74
    in Montana lignite, 216t
    Victorian brown coal, 8,11f
Calcium acetate
    phase-shift subtracted Fourier
        transform, 169f
    X-ray absorption near-edge
        spectra, 168f
Calcium-containing coals, X-ray
    absorption spectra, 161
Calcium-enriched macerals, backscat-
    tered electron intensity, 166
Calcium oxide, 256
Calcium-rich phases, computer-con-
    trolled SEM, 161
Carbon
    Beulah lignite, 89t
    burning rate, equation, 244
    northern Great Plains lignites, 40t
    in Victorian brown coal, 10f
Carbon dioxide adsorption, 89
Carbon monoxide influence,
    devolatilization, 237
Carbon preference indices
    n-alkanes vs. depth, 127f
    calculation, 110
    n-fatty acids vs. depth, 127f
Carbonate formation, Victorian brown
    coal, 13
Carboxyl groups
    decomposition, 222,223f
    Wilcox lignite, 134,140t
Carboxylate concentration, northern
    Great Plains lignites, 43
Cation(s)
    effects on weight loss, 220t
    exchangeable, role in lignite
        pyrolysis, 213-25
    selection, 260
Cation exchange, conditions, 217t
Cation loading, 217
CCSEM--See Computer-controlled scan-
    ning electron microscopy
Cellulose, 294-95
    coalification, 299f
Cesium, position in lignite seam, 188

Char(s), 138,271f
  combustion
    rate data, 250f
    reactivity, 243-54
  kinetic data, combustion, 249t
  preparation, 268
Char gasification, calcium
  catalyst, 268
Chemical structure, Rhenish brown
  coals, 15-38
Clay-rich sediments, 191
Coal-ash chemistry, Victorian brown
  coal, 7t
Coal seam, variations, 180
Coexchanged chars
  metal contents, 270t
  reactivities, 272
Coke combustion
  kinetic data, 249t
  rate data, 250f
Coking, Rhenish brown coals, 28-30
Color index
  See also Lithotype
  Victorian brown coal, 7
Colstrip
  computer-controlled SEM
    results, 164t
  Mossbauer results, 162t
Combustion
  aluminum oxide, role, 14
  cation effects, 255-66
  particle structure effect, 245
  reactivity, of chars, 243-54
  reactor system, 260
  stages, 266
  Texas lignite, 264
Computer-controlled scanning electron
  microscopy
  bituminous coals, 167t
  fouling deposits, 173t
  principles, 160
  results, 164t
CPI--See Carbon preference indices
Cross polarization-magic angle spinning
  Victorian brown coal, 111
  Wilcox lignite, 136f
Cyclic hydrocarbons, 153-57

                         D

Decarboxylation, kinetic
  parameters, 222t
Deno oxidation--See Pertrifluoroacetic
  acid oxidation
Deposit, formation, 256
Depositional factors, influences, 181
Depth, chemical variation on Victorian
  brown coal, 109-31

Devolatilization
  CO influence, 237,240f
  rate equation, 236
Diacids, composition in
  lithotypes, 122t
Diaryl methane units in
  lithotypes, 117
Dibenzyl ether reactions, iron-based
  catalyst, 280
Diols, 123
Diphenylmethane, bond cleavage, half-
  life, 43
Dry brown coal, microstructure
  determination, 95-108

                         E

Elemental distribution,
  interpretation, 188
Entrained flow
  reactor, 214,215f,260,261f
  release of tars, 221t
  tubular, 247f
  weight loss, 219f
Entrained gasification, 73
Ether(s)
  conversion, cation effects, 281f
  reaction with Victorian brown
    coal, 280
Ether links, liquefaction, 298
Eveleth, computer-controlled SEM, 164t
EXAFS, principles, 160
Exchangeable cations
  effects on tar release, 221-22
  role in lignite behavior, 213-25,267
Extended X-ray absorption fine
  structure, principles, 160
Extraction, 147
Extracts, GC profiling, 157

                         F

n-Fatty acids
  vs. depth, carbon preference
    indices, 127f
  in lithotypes, 128
Field variations, Victorian brown
  coal, 12t
Fixed carbon
  Beulah lignite, 89t
  northern Great Plains lignites, 40t
  Texas lignite, 60-61t
Fouling
  analysis, 170,173
  deposits, computer-controlled
    SEM, 173t

Fourier transform
  phase-shift subtracted, 170
  techniques, 160
Fractions, characterization, 123,130
Fuel oil, use, 255

G

Gas adsorption, brown coal, 104t
Gas chromatography, hydrocarbon
  extracts, 145-58
Gas chromatography/mass
  spectrometry, 145-58
Gas chromatography profiling,
  extracts, 157
Gasification
  entrained, 73
  Rhenish brown coals, 30-31
  Texas lignite, 71-72
  underground, 73
GC/MS--See Gas chromatography/mass
  spectrometry
Geochemical variations, among
  seams, 180,191
German coals--See Rhenish brown coals
Gulf Coast lignites, trace
  elements, 62
Gyration, mean radius, equation, 101

H

Heating value
  from hydropyrolysis of Wyoming
    coal, 234f
  Rhenish brown coals, 23,24f,25f
  Texas lignite, 288
Helium, in entrained flow reactor, 216
Heterogeneous combustion, ion-exchan-
  geable cations, 257
Hexane-soluble extracts, 147t
High-temperature Winkler
  gasification, 30
Highvale
  acyclic hydrocarbons, 149t
  analyses, 146t
  chloroform extracts, 145-58
  hexane-soluble extracts, 150t
HKV--See Hydrogasification of lignite
Homohopane, ratio of
  diastereomers, 129f
Homophane, structure, 128
Hopanoids, biologically derived, 128
HTW--See High-temperature Winkler
  gasification
Humic acid, solvent
  extractable, 123-24
Huminite content, Rhenish brown
  coals, 21,22f

Humotelite content, Rhenish brown
  coals, 28
Hydrocarbon(s)
  extracts, 145-58
  production, 289
Hydrogasification of lignite, 30
Hydrogen
  Beulah lignite, 89t
  northern Great Plains lignites, 40t
  Rhenish brown coals, 23,28
  variation in Victorian brown
    coal, 10f
Hydrogen donor solvent, effect on
  lignite dissolution, 292
Hydroliquefaction
  iron-tin catalysts, 277
  radical propagation
    reactions, 281-82
  Victorian brown coal, 275-86,279t
    conversion vs. pressure, 278f
    conversion vs. temperature, 278f
    iron effects, 282t
    mechanism, 284f
    reaction temperature effects, 277
Hydropyrolysis, 227-42
  heating rate influences, 230
  kinetic parameters, 238t
  thermograms, 232f

I

Illinois coal
  analysis, 228t
  hydropyrolysis parameters, 238t
  pyrolysis with Syngas, 241t
Inorganic affinity, 188
Inorganic constituents
  analysis, 159-74
  geochemical variation, North Dakota
    lignite, 175-92
  Sentinel Butte formation, 177t
Ion exchange conditions, 217t
Ion exchangeable cations, 257
Ion profiles, 153
  Wyodak subbituminous coal, 155f
IR spectroscopy, absorption mode, 114
Iron
  activity, 284
  detection method, 160
  effect on ether conversions, 281f
  effects on preasphaltene
    hydrogenation, 282t
Iron oxyhydroxide, Mossbauer
  spectra, 163f
Iron-tin catalyst, 279
  in hydroliquefaction, 277
Isoprenoids, liquefaction
  products, 149t

J

Jarosite, Mossbauer spectra, 163f

K

Kaolinite, 62
  in Montana lignite, 217t
Kentucky coal
  analysis, 228t
  hydropyrolysis parameters, 238t
  pyrolysis with Syngas, 241t
Kinneman Creek Bed lignite, 178f,179t
  distribution
    aluminum, 186f
    calcium, 189f
    nickel, 186f
    strontium, 189f
    vanadium, 187f
Kratky camera, 96,97f,98

L

Latrobe Valley, Australia, 3
  See also Australian subbituminous
      coals
Lignite
  Beulah
    See Beulah lignite
  char gasification, catalysis, 267-74
  char reactivity, decreasing, 267
  combustion, cation effects, 255-66
  definition, 41
  dissolution, hydrogen donor solvent
      effects, 292
  Gulf Coast, trace elements, 62
  hydrogasification, 30
  Kinneman Creek bed
    See Kinneman Creek bed lignite
  Montana
    See Montana lignite
  North Dakota
    See North Dakota lignite
  northern Great Plains
    See Northern Great Plains
        lignite
  production in 1990, 53
  Pust
    See Pust lignite
  pyrolysis
    cation effects, 255-66
    rapid, exchangeable
        cations, 213-25
  scattering curve, 82f,85f
  specific surfaces, 88
  structure, 300
  Texas, 53-75
    See Texas lignite and Wilcox
        lignite

Lignite--Continued
  Wilcox
    See Texas lignite and Wilcox
        lignite
  X-ray specific surfaces, 87t
Liptinite, in Rhenish brown
    coals, 21,22f
Liquefaction
  ether links, 298
  Rhenish brown coals, 31,33
  Texas lignite, 73
  volatile species, 298
Lithotype
  See also Color index
  brown coal, heating value, 24f
  compositon, 122t
  definition, 109
  diaryl methane units, 117
  distinguishing characteristics, 109
  pertrifluoroacetic acid oxidation,
      product distribution, 119t
  solvent extraction, 110
  Victorian brown coal, 8,10f
    chemical variation, 109-31
Low silica slags, Urbain
    correlations, 208
Low-temperature ashing, 59,62,180
  minerals determined, 181t
  of Montana lignite, 217t
LTA--See Low-temperature ashing
Lupolen, scattering curve, 100f

M

Macerals, Ca enriched, backscattered
    electron intensity, 166
Macropores
  distinguishing specific surfaces, 89
  size, 83
Magnesium
  and calcium, char gasification
      catalyst, 270
  cation effects, 218
  coexchanged chars, 270t
  effects
    on combustion, 264,265f
    on pyrolysis, 218,262
    on Texas lignite, weight
        loss, 263f
  lignite char gasification, 267-74
  in Montana lignite, 216t
  Victorian brown coal, 11f
Magnesium oxide, 272
Malonic acid, derivation, 117
Methane-soluble extracts, 147t
Micropores
  size, 83
  volume, SAXS, 101
  X-ray specific surface, 89

Microstructure determination, small-
    angle X-ray scattering, 95-108
Millmerran char, 246,251
    combustion, 252f
        rate data, 250f
    kinetic data, 249t
    particle temperature effects, 248f
Minerals
    brown coal, Victorian, 6t
    group components, 5
Moisture
    Beulah lignite, 89t
    northern Great Plains lignites, 40t
    Texas lignite, 59,60-61t,62
Monoacids, composition in
    lithotypes, 122t
Montana lignite
    cation contents, 216t
    chars
        preparation, 268
        reactivities, 269f
    mineral matter composition, 217t
    weight loss, 218,219f
        from pyrolysis, 224f,225f
Morwell
    analyses, 146t
    chloroform extracts, 145-58
    hexane-soluble extracts, 150t
Mossbauer spectra
    bituminous coals, 167t
    boiler-wall slag deposits, 172f
    results, 162t
    use, 160

N

NAA--See Neutron activation analysis
Neutrals fraction, 117,123
    chromatographic separation, 120f
Neutron activation analysis, 179
    results, 182-85t
New Zealand bituminous coal char,
    kinetic data, 249t
Newton's law of viscosity, 197-98
NGPL--See Northern Great Plains lig-
    nites
Nickel distribution, Kinneman Creek
    Bed lignite, 186f
Nickel-molybdenum catalyst, 277,285
Nitrogen
    Beulah lignite, 89t
    northern Great Plains lignites, 40t
Nonpyritic iron, Victorian brown
    coal, 11f
North Dakota lignite
    analysis, 228t
    computer-controlled SEM, 164t
    fouling deposits, 170

North Dakota lignite--Continued
    geochemical variation of inorganic
        constituents, 175-92
    hydropyrolysis parameters, 238t
    Mossbauer results, 162t
    pyrolysis with Syngas, 241t
    samples, 179t
Northern Great Plains lignites, 39-50
    analyses, 40t
    aromaticity, 43
    ash composition, 41t
    vs. bituminous coals, 40-41
    carboxylate concentration, 43
    organic structural relationships, 41
    vitrinite concentrates, 43

O

Oblate ellipsoids, 106f
Oleanane, structure, 156
Overburden-underclay samples, minor
    element concentrations, 192t
Oxygen
    Beulah lignite, 89t
    northern Great Plains lignites, 40t
Oxygen functional groups, Victorian
    brown coal, 7

P

PAH--See Polycyclic aromatic hydrocar-
    bons
Particle-dimension distribution,
    equation, 91
PDSC--See Pressure differential scan-
    ning calorimetry
Pertrifluoroacetic acid oxidation, 117
    brown coal lithotypes, 118t
    product distribution in
        lithotypes, 119t
    proton yield, 118f
    Wilcox lignite, 135
Petrographical properties, Rhenish
    brown coals, 15-38
Petroleum coke, kinetic data, 249t
Phase-shift subtracted Fourier
    transform, 170
Phenolic groups, Wilcox lignite, 140t
Plastic, viscosity
    characteristics, 198
Polycyclic aromatic hydrocarbons,
    brown coal, 128,130f
Pore
    classification, 83,96
    dimension distributions, 90-93,103
    formation, 95
Potassium, in Montana lignite, 216t

Powhatan
  acyclic hydrocarbons, 149t
  n-alkane extract distribution, 152f
  analyses, 146t
  chloroform extracts, 145-58
  hexane-soluble extracts, 150t
Precipitator ash, aluminum effects, 14
Pressure differential scanning
    calorimetry, 41-43
Pressure pyrolysis apparatus, 229f
Prolate ellipsoids, 105,106f
Pseudoplastic, viscosity,
    characteristics, 198
Pust lignite
  backscattered electron image, 165f
  computer-controlled SEM
      results, 164t
  Mossbauer results, 162t,163f
  phase-shift subtracted Fourier
      transform, 169f
  X-ray absorption near-edge
      spectra, 168f
Pyrite, Mossbauer spectra, 163f
Pyrolysis
  cation effects, 218,255-66
  chemical expression, 236
  heating rate effects, 235f
  hydrogen effects, 230
  kinetics, 236
  pressure, apparatus, 229f
  rate-determining step, 138
  with Syngas, kinetic
      parameters, 241t
  Texas lignite, 262
  in wet nitrogen and carbon
      dioxide, 224
  Wyoming coal, FTIR monitor, 235f

                        Q

Quality
  lignites, 57
  vs. refining behavior, Rhenish brown
      coals, 28-30
Quartz, in Montana lignite, 217t

                        R

Radical propagation reactions,
    hydroliquefaction, 281-82
Rhenish brown coals, 15-38
  ash content, 21
  briquetting, 28-30
  characterization, 16,19f,20f,21
  chemo-physical characterization, 24f
  coking, 28-30
  composition, 16-21

Rhenish brown coals--Continued
  direct hydrogenation, 33
  gasification, 30-31
  heating value, 23,24f
    vs. hydrogen, 25f
  huminite content, 21,22f,25f
  humotelinite vs. lipto-
      humodetrite, 27f
  humotelite content, 28
  hydrogasification, conversion, 32f
  hydrogen content, 23,28
  hydrogenation, 31,32f,34f,35
  liptinite content, 21,22f,25f
  liquefaction, 31,33
  oxygen functional groups,
      distribution, 18f
  quality vs. refining behavior, 28-30
  structure, 16-21,17f
  volatiles, 23,26f
Rock-Eval analysis, 114
  Victorian brown coal, 111,116f
Rosebud
  computer-controlled SEM
      results, 164t
  Mossbauer results, 162t
Rotator method, 99

                        S

S^2 correlations, 201
  failures, 200
Sampling, 177
SAXS--See Small-angle X-ray scattering
Scattered intensity, equation, 80
Scattering curve
  corrected, 102f
    for lignite, 85f
  least-squares fit, 83
  lignite, 82f
  sample preparation, 81
  three dimensional, 90
Scattering power, rotator method, 99
Seam(s), variation in brown coals, 7-8
Seam distribution, and geochemical
    properties, 191
Semianthracite, kinetic data, 249t
Sentinel Butte formation, inorganic
    constituents, 177t
Sesquiterpenes, detection, 153
Silica, Texas lignite, 62
Slag(s)
  composition data, 202-3t
  formation, 256
  Urbain correlation, 204
  viscosity vs. temperature, 199f
Slag viscosity
  apparatus, 196
  coal pulverization, 196

Slag viscosity--Continued
  composition correlations, 200-208
  hysteresis effect, 198
  measurement and prediction, 195-209
Slagging
  behavior, important compounds, 173
  deposits, analysis, 170,173
Small-angle X-ray scattering
  absolute intensity
    measurement, 98-99
  brown coal, 104t
  data analysis, 79-90
  mean radius of gyration, 101
  micropore volume, 101
  microstructure determination, 95-108
  sample preparation, 81,98
  scattering intensity
    determination, 80
    zero angle, 103
  submicroscopic porosity, 79-94
  system, 82f
  transmission factor, 99
  use, 79
Sodium, 257
  boiler fouling, 13
  cation effects, 218
  effects on pyrolysis, 218
  in Montana lignite, 216t
  Victorian brown coal, 11f
Solvent extraction, lithotypes, 110
Soxhlet extraction, 148f
Specific surface, 105
Steranes, detection, 153
Strontium
  in Kinneman Creek Bed, 189f
  in Montana lignite, 216t
Structure, 300f
  dimensional measurements, 80
  Rhenish brown coals, 16-21
Subbituminous coal
  Australian
    chars, properties, 244,246t
    combustion reactivity, 243-54
  specific surfaces, 88
  X-ray specific surfaces, 87t
Submicroscopic porosity, small-angle
  X-ray scattering, 79-94
Succinic acid, derivation, 117
Sulfided Ni-Mo catalyst, 285
Sulfur
  Beulah lignite, 89t
  northern Great Plains lignites, 40t
  Texas lignite, 60-61t,67t
Sulfur retention, ash, 68,200
Surface
  proportionality, 92
  X-ray specific, 87t

T
Tar release, exchangeable cations
  effects, 221-22
Tar samples, analysis, 222
Technological behavior, Rhenish brown
  coals, 15-38
Temperature programmed pyrolysis--See
  Rock-Eval analysis
Texas lignite, 53-75
  See also Wilcox lignite
  alkane fractions, 295f
  analyses, 60-61t,259f
  ash composition, 64-65t
  atmospheric fluidized-bed
    combustion, 70-71
  boiler fouling, 69
  combustion, 69-70,264
    weight loss, 265f
  continuity, 57
  cyclic deposition, 55f
  deep-basin, 58f
  desulfurization, 68
  direct combustion, 69
  economic studies, 73
  gasification, 68,71-72
  heating values, 288
  liquefaction, 68,73,287-300
    feed gas effect, 296f
    volatile species, 298
  major oxides, 62-66
  mineral matter, 59,62
  near-surface, 56f
  near-term utilization, 73
  oxidation, 66
  petrography, 59
  properties, 59
  pyrolysis, 66,262
  quality, 57
  seam thickness, 57
  silica, 62
  stratigraphy, 54,57
  structure, 294-300
  sulfur forms, 67t
  THF solubles, 293f,295f
  trace elements, 62-66
  uranium, 63
  utilization, 69
  washability characteristics, 66
  weight loss, 263f
Thermogravimetry, apparatus, 229f
Thixotropic, viscosity
  characteristics, 198
Three-dimensional scattering, 90
Tin catalysis
  ether conversions, 281f
  preasphaltene hydrogenation, 282t
  hydroliquefaction, 277

Total acidity
    determination, 134-35
    Wilcox lignite, 140t
Total weight loss, kinetics, 221,221t
Trace elements, Texas lignite, 62-66
Transitional pores, 83,89
Triacids, composition in
    lithotypes, 122t
Tricyclic terpenoids, detection, 153
Tubular entrainment flow reactor, 247f

                         U

UCG--See Underground coal gasification
Underclay, determination, 180
Underground coal gasification, 71-73
Uranium, 188
    Texas lignite, 63
Urbain correlations
    equation, 204
    low-silica slags, 208
    slags, 204

                         V

Vanadium
    in Kinneman Creek Bed, 187f
    position in seam, 188
Victorian brown coal
    See also Australian subbituminous
        coals
    n-alkane distribution, 124
    aromaticity vs. depth, 113f
    ash content vs. minerals, 6t
    atomic absorption, 4-5
    bed-moisture content, 6t
    calcium, 11f
    carbon variation, 10f
    carbonate formation, 13
    characteristics, 3-14,109-31
    coal field variations, 8-12
    color index, 7
        CP-MAS, 111
    extractable polycyclic aromatic
        hydrocarbons, 130f
    fatty acids, 128
    hydrogen index, 116f
    hydrogen variation, 10f
    hydroliquefaction, 275-86,279t
        effects, 278f
        mechanism, 284f
        temperature effects, 277,278f
    inorganic content, 4-5
    IR absorption coefficients vs.
        depth, 115f
    lithotypes, 8,10f

Victorian brown coal--Continued
    magnesium, 11f
    mineral content, 4-5
    moisture content, 4
    Morwell
        Ca content, 8
        coal-ash chemistry, 7t
        depth variations, 9f
    nonpyritic iron, 11f
    oxygen functional groups, 7
    oxygen index, 116f
    pertrifluoroacetic acid
        oxidation, 111
    preasphaltene hydrogenation, 282t
    product distribution, 119t,279t
    properties, 4-8
        field variations, 12t
    reaction with ether compounds, 280
    Rock-Eval pyrolysis, 111,116f
    seams, variations, 7-8
    sodium, 11f
    water in clays, 5
    X-ray diffraction, 13
    X-ray fluorescence, 4-5
Viscosity
    equation, Arrhenius form, 201
    Newton's law, 197-98
Vitrinite concentrates, northern Great
    Plains lignites, 43
Volatiles
    Beulah lignite, 89t
    northern Great Plains lignites, 40t
    Rhenish brown coals, 23
    Texas lignite, 60-61t

                         W

Wandoan char, 251
    combustion behavior, 252f
    kinetic data, 249t
Washability characteristics, Texas
    lignite, 66
Watt-Fereday correlations, 200,201
Weight loss
    cation effects, 220t
    determination, 260
    total, kinetics, 221,221t
Wet brown coal
    microstructure determination, 95-108
    small-angle X-ray scattering, sample
        preparation, 98
Whole coals
    aromaticity vs. depth, 113f
    bond equivalence diagram, 112f
    characterization, 111-23
Wilcox lignite
    See also Texas lignite
    acid groups,
        determination, 134-35,138

Wilcox lignite--Continued
    aliphatic chains, 141
    alkyl phenols, 142t
    analyses, 137t
    aromaticity, 135
    carboxyl group concentration, 134
    CP-MAS spectra, 136f
    dimethyl sulfoxide extract, 137t
    liquefaction, 135,142t
    organic constituents, 143f
    oxidation products, 141
    pertrifluoroacetic acid
        oxidation, 135,140t
    proximate analyses, 134
    quality, 57
    structural features, 133-44,139t
Wyodak
    acyclic hydrocarbons, 149t
    n-alkane extract distribution, 151f
    analyses, 146t
    chloroform extracts, 145-58
    chromatograms, 155f
    hexane-soluble extracts, 150t
    sesquiterpene distribution, 153
Wyoming coal
    analysis, 228t
    hydropyrolysis
        heating rate effect, 234f
        parameters, 238t
    pyrolysis
        FTIR monitor, 235f

Wyoming coal--Continued
    Syngas, 241t
    thermograms, 233f
    TGA thermograms, 231f

X

X-Ray absorption spectra
    Ca-containing coals, 161
    near-edge, 161,171f
X-Ray diffraction, 180
    minerals determined, 181t
    Victorian brown coal, 13
X-Ray fluorescence
    Victorian brown coal, 4-5
    180,182-85t
X-Ray scattering, 80
X-Ray specific surface, 87t
    Beulah lignite, 88
    micropores, 89
XANES--See X-Ray absorption spectra,
    near-edge
XRD--See X-Ray diffraction
XRF--See X-Ray fluorescence

Y

Yallourn brown coal char, kinetic
    data, 249t

Indexing and production by Deborah Corson
Jacket design by Pamela Lewis

Elements typeset by Hot Type Ltd., Washington, D.C.
Printed and bound by Maple Press Co., York, Pa.

RECENT ACS BOOKS

"Computers in Flavor and Fragrance Research"
Edited by Craig B. Warren and John P. Walradt
ACS SYMPOSIUM SERIES 261; 157 pp.; ISBN 0-8412-0861-1

"Polymers for Fibers and Elastomers"
Edited by Jett C. Arthur, Jr.
ACS SYMPOSIUM SERIES 260; 422 pp.; ISBN 0-8412-0859-X

"Treatment and Disposal of Pesticide Wastes"
Edited by James N. Seiber and Raymond F. Krueger
ACS SYMPOSIUM SERIES 259; 384 pp.; ISBN 0-8412-0858-1

"Stable Isotopes in Nutrition"
Edited by Judith R. Turnlund and Phyllis E. Johnson
ACS SYMPOSIUM SERIES 258; 240 pp.; ISBN 0-8412-0855-7

"Bioregulators: Chemistry and Uses"
Edited by Robert L. Ory and Falk R. Rittig
ACS SYMPOSIUM SERIES 257; 286 pp.; ISBN 0-8412-0853-0

"Polymeric Materials and Artificial Organs"
Edited by Charles G. Gebelein
ACS SYMPOSIUM SERIES 256; 208 pp.; ISBN 0-8412-0854-9

"Pesticide Synthesis Through Rational Approaches"
Edited by Philip S. Magee, Gustave K. Kohn, and Julius J. Menn
ACS SYMPOSIUM SERIES 255; 351 pp.; ISBN 0-8412-0852-2

"Advances in Pesticide Formulation Technology"
Edited by Herbert B. Scher
ACS SYMPOSIUM SERIES 254; 264 pp.; ISBN 0-8412-0840-9

"Structure/Performance Relationships in Surfactants"
Edited by Milton J. Rosen
ACS SYMPOSIUM SERIES 253; 356 pp.; ISBN 0-8412-0839-5

"Chemistry and Characterization of Coal Macerals"
Edited by Randall E. Winans and John C. Crelling
ACS SYMPOSIUM SERIES 252; 192 pp.; ISBN 0-8412-0838-7

"Conformationally Directed Drug Design:
Peptides and Nucleic Acids as Templates or Targets"
Edited by Julius A. Vida and Maxwell Gordon
ACS SYMPOSIUM SERIES 251; 288 pp.; ISBN 0-8412-0836-0

"Polymer Blends and Composites in Multiphase Systems"
Edited by C. D. Han
ADVANCES IN CHEMISTRY SERIES 206; 400 pp.; ISBN 0-8412-0783-6

"The Chemistry of Solid Wood"
Edited by Roger M. Rowell
ADVANCES IN CHEMISTRY SERIES 207; 588 pp.; ISBN 0-8412-0796-8